ESTABLISHING A CGMP LABORATORY AUDIT SYSTEM

ESTABLISHING A CGMP LABORATORY AUDIT SYSTEM
A Practical Guide

David M. Bliesner
Delphi Analytical Services, Inc.
Indian Rocks Beach, Florida

WILEY-INTERSCIENCE

A JOHN WILEY & SONS, INC., PUBLICATION

Published by John Wiley & Sons, Inc., Hoboken, New Jersey
Published simultaneously in Canada

For general information on our other products and services or for technical support, please contact
our Customer Care Department within the United States at 877-762-2974, outside the United States
at 317-572-3993 or fax 317-572-4002.

Wiley also publishes its books in a variety of electronic formats. Some content that appears in print
may not be available in electronic formats. For more information about Wiley products, visit our web
site at www.wiley.com.

Library of Congress Cataloging-in-Publication Data:

Bliesner, David M.
    Establishing a CGMP laboratory audit system : a practical guide / David
M. Bliesner.
       p. cm.
    Includes bibliographical references and index.
    ISBN-13 978-0-471-73840-4
    ISBN-10 0-471-73840-9 (cloth : alk. paper)
    1. Pharmaceutical industry—Law and legislation—United States.　I. Title.
    [DNLM:　1. Drug Industry—standards—United States.　2. Laboratories—standards—
United States.　3. Management Audit—methods—United States.　4. Drug Industry—legislation &
jurisprudence—United States.　5. Laboratories—legislation & jurisprudence—United States.
QV 736 B648c 2006]
KF1879.B55 2006
343.7307′86151—dc22

2005024178

Printed in the United States of America

10 9 8 7 6 5 4 3 2 1

To my wife Kathy, and my children Nick, Sam and Erin
for their love, support and patience.

CONTENTS

PREFACE

Delphi Analytical Services, Inc. has spent the last several years helping companies in the pharmaceutical industry improve their level of compliance with current good manufacturing practices (CGMPs). This involvement has included large and small companies who have already been subject to regulatory action from the U.S. Food and Drug Administration (FDA) as well as companies who are taking preventative measures to avoid regulatory action. As part of this effort, a significant amount of time has been spent reviewing the quality systems associated with analytical laboratories.

The FDA mandates that a drug firm and its laboratory be operated in a state of control by employing conditions and practices that assure compliance with the intent of the Federal Food, Drug, and Cosmetic Act and portions of the CGMP regulations that pertain to it. Specifically, a laboratory, which is in a state of control, provides services that confirm the company is producing finished drug products of sufficient quality, known strength, proper identity, and known purity.

In order to demonstrate that your firm is in control, data are need to support your position. These data are obtained by executing a well organized and systematic laboratory audit.

In addition to demonstrating current control, you must show that you will be in control in the future. Therefore, you must also demonstrate you have a system in place to continually monitor the status of compliance within your laboratory and correct deficiencies if they are discovered.

Establishing a CGMP Laboratory Audit System: A Practical Guide is a systematic approach for auditing your laboratory to demonstrate to your organization and, ultimately, to the FDA, that you are in control of your laboratory system. In addition, this guide helps you accomplish the goal of establishing sustainable compliance within your laboratory. This text is a "how to" book—how to establish a current good manufacturing practices (CGMP) laboratory audit system. The intended purpose of the book is to instruct through detailed flowcharts, checklists, and descriptions, the process of establishing a CGMP laboratory audit system from scratch or to upgrade existing systems to comply with current industry practices. Moreover, this process is an excellent means to teach or refresh laboratory personnel on the nuances of operating a modern pharmaceutical laboratory under CGMPs.

Specifically designed for laboratories regulated by the U.S. government, this guide is useful for:

- Facilities operating under current good manufacturing practices (CGMPs)
- Facilities operating under current good laboratory practices (CGLPs)
- Facilities operating under ISO standards.

However, any laboratory can benefit from the level of control obtained by the guide and the corresponding incremental gains in efficiency and productivity from implementing such a system.

This guide is not an academic treatise, but a collection of real-world tools, that can be applied immediately and directly to your laboratory. Some unique and special features presented include:

- Detailed audit checklists corresponding to the seven subelements which compose the laboratory control system
- A real-world audit summary report example template
- The current FDA guidance document on the subject of drug manufacturer inspections
- Audit tools and templates, such as suggested meeting agendas, audit routines, audit calendars, and data capture forms.

All of these tools and others are provided on a CD-ROM, which accompanies this book, for easy application by the end users in their own laboratories. Moreover, these tools and templates are provided in readily modifiable formats so that they maybe tailored to fit the needs of the individual organization. The inclusion of these practical tools makes this guide unique. It would require untold personnel hours to develop these checklists and example templates individually. In fact in smaller organizations, the time, talent, and experience to create such tools is most likely outside their capabilities.

To my knowledge no such detailed instructional text for implementing CGMP laboratory quality systems (including detailed example templates of critical end-user documents) exists in the marketplace. I hope you find *Establishing a CGMP Laboratory Audit System: A Practical Guide* useful and wish you the best in your continuing quest to attain compliance and improve quality.

DAVID M. BLIESNER

Indian Rocks Beach, Florida
February 2006

CHAPTER 1

INTRODUCTION TO THE QUALITY SYSTEMS APPROACH TO CGMP COMPLIANCE

1.1 OVERVIEW OF QUALITY SYSTEMS

The Food and Drug Administration (FDA) mandates that a drug firm, and therefore its laboratory, be operated in a state of control by employing conditions and practices that assure compliance with the intent of The Federal Food, Drug, and Cosmetic Act and portions of the Current Good Manufacturing Practice (CGMP) regulations (e.g., 21 CFR Parts 210 and 211) that pertain to it. Activities found in drug firms, including operation of the laboratory, can be organized into systems that are sets of operations and related activities. Control of all systems helps to ensure the firm produces drugs that are safe, have the proper identity and strength, and meet the quality and purity characteristics as intended.

For drug firms, the FDA has outlined the following general scheme of systems that impact the manufacture of drugs and drug products:

1. *Quality System.* This system assures overall compliance with CGMPs and internal procedures and specifications. The system includes the quality control unit and all of its review and approval duties (e.g., change control, reprocessing, batch release, annual record review, validation

Establishing a CGMP Laboratory Audit System. By David M. Bliesner
Copyright © 2006 John Wiley & Sons, Inc.

protocols, and reports, etc.). It includes all product defect evaluations and evaluation of returned and salvaged drug products. (See CGMP regulation 21 CFR 211 Subparts B, E, F, G, I, J, and K.)

2. *Facilities and Equipment System.* This system includes measures and activities that provide an appropriate physical environment and resources used in the production of the drugs or drug products, including:

   (a) Buildings and facilities with maintenance;

   (b) Equipment qualifications (installation and operation), equipment calibration and preventative maintenance, and cleaning and validation of cleaning processes, as appropriate. Process performance qualifications are included as part of process validation done within the system where the process is employed and;

   (c) Utilities that are not intended to be incorporated into the product such as HVAC, compressed gases, steam, and water systems.

   (See CGMP regulation 21 CFR 211 Subparts B, C, D, and J.)

3. *Materials System.* This system includes measures and activities to control finished products and components including water or gases that are incorporated into the product, containers, and closures. It includes validation of computerized inventory control processes, drug storage, distribution controls, and records. (See CGMP regulation 21 CFR 211 Subparts B, E, H, and J.)

4. *Production System.* This system includes measures and activities to control the manufacture of drugs and drug products including batch compounding, dosage form production, in-process sampling and testing, and process validation. It also includes establishing, following, and documenting performance of approved manufacturing procedures. (See CGMP regulation 21 CFR 211 Subparts B, F, and J.)

5. *Packaging and Labeling System.* This system includes measures and activities that control the packaging and labeling of drugs and drug products. It includes written procedures, label examination and usage, label storage and issuance, packaging and labeling operations controls, and validation of these operations. (See CGMP regulation 21 CFR 211 Subparts B, G, and J.)

6. *Laboratory Control System.* This system includes measures and activities related to laboratory procedures, testing, analytical methodology development, validation or qualification/verification, and the stability program. (See CGMP regulation 21 CFR 211 Subparts B, I, J, and K.)

As stated in (6) above, FDA considers a firm's laboratory control system to be a key element in CGMP compliance. Within the laboratory control systems are at least seven additional subsystems or subelements which include:

- Laboratory managerial and administrative systems,
- Laboratory documentation practices and standard operating procedures,
- Laboratory equipment qualification and calibration,
- Laboratory facilities,
- Methods validation and technology transfer,
- Laboratory computer systems, and
- Laboratory investigations.

Establishing and maintaining quality systems and subsystems demonstrates control.

1.2   QUALITY SYSTEMS AND COMPLIANCE WITH CGMPs: REASONS FOR AUDITING YOUR LABORATORY

The purpose for auditing your laboratory is to demonstrate to your organization and ultimately to FDA that you are in control of your laboratory control system.

In order to demonstrate control, data is needed to support your position. These data are obtained by executing a well-organized and systematic laboratory audit.

In addition to demonstrating current control, you must show future control. Therefore, you must also have in place a system that to continually monitors the status of compliance within laboratory and corrects deficiencies if discovered.

1.3   GOALS OF AUDITING YOUR LABORATORY

In short, the goals of a laboratory audit are:

- Demonstrate control by conducting the audit and generating data to support your position.
- If not in control then:
  - Show that you know why you are not in control;
  - Show that you know which areas are out of compliance;
  - Show that you know which areas have the greatest impact;
  - Develop interim controls to mitigate the impact of the areas with the greatest risk;
  - Develop a plan to put you back in control;

  ○ Implement the plan; and
  ○ Generate a system to continually monitor your state of compliance so you stay in control in the future (e.g., sustainable compliance).

1.4   LABORATORY AUDIT PHASES

As stated in the preceding list, a well-organized and systematic laboratory audit must be executed in order to obtain data to prove control. To accomplish this, the audit may be organized into the following phases:

- Preparation phase,
- Audit and data capture phase,
- Reporting phase,
- Corrective action phase,
- Verification phase, and
- Monitoring phase.

Details of the design and implementation for each phase are described in the remaining chapters of this book. In addition, some of the tools, templates, and examples needed to complete such an audit are included in the Appendices.

1.5   INTEGRATION WITH EXISTING PROGRAMS

One of the strengths of the laboratory control system audit process described in this guide is that it allows for easy integration and linkage with existing audit programs and data. Specifically:

- Data collected from previous internal audits, 483 observations, external audits, and gap analyses are linked and compiled via use of the laboratory audit form (LAF) data capture instrument.
- Existing corrective action project plans become part of the corrective action phase of this process and are managed as one coherent effort.

1.6   MODIFIABLE AND SCALABLE APPROACH

In addition to the ability to integrate this approach into existing systems, the guide is also constructed with the following major characteristics:

- *Scalable.* The audit approach described here is useful regardless of the size of the facility. It works whether your organization has 10, 100, or several hundred employees. Simply scale the magnitude of the audit based on the availability of resources at your facility and match those laboratories that constitute your quality operations.

- *Modifiable.* The tools and templates outlined in this book are designed not only to instruct but to be copied and modified. Take them and modify them a little or modify them a lot. They are meant to save time and prevent reinvention the wheel.

REFERENCE

1. Food and Drug Administration, *Compliance Program Guidance Manual For FDA Staff*, "Drug Manufacturing Inspections Program," 7356.002, February 2001.

BIBLIOGRAPHY

Food and Drug Administration, Code of Federal Regulations, Food and Drugs, Title 21 Parts 210 and 211 "Current Good Manufacturing Practice in Manufacturing, Processing, Packing, or Holding of Drugs: General" and "Current Good Manufacturing Practice for Finished Pharmaceutical," Revised April 1, 2005.

FDA Guidance for Industry: *Analytical Procedures and Methods Validation*, draft, August 2000.

Food and Drug Administration, *Compliance Program Guidance Manual For FDA Staff*, "Inspections of Licensed Biological Therapeutic Drug Products," 7356.002M, October 2003.

HHS Publication, *Medical Device Quality Systems Manual: A Small Entity Compliance Guide*, FDA 97-4179, December, 1996.

ICH Q2A, *Text on Validation of Analytical Procedures*, March 1995.

ICH Q2B, *Validation of Analytical Procedures: Methodology*, May 1997.

US Pharmacopoeia—National Formulary, United States Pharmacopoeia Convention, Inc., Rockville, MD, 2005.

CHAPTER 2

PREPARING FOR THE AUDIT

2.1 PROCEDURE

The key to executing a well-organized and systematic laboratory audit is taking the time to develop the proper audit team organizational structure, define work functions, assign roles and responsibilities, conduct audit familiarization and overview sessions, and perform audit team training. The steps in this process are shown in Figure 2.1 and described in Table 2.1.

Some details for each step are summarized in Table 2.1.

2.2 AUDIT TOOLS AND TEMPLATES

As referenced in Step 7 of Table 2.1 (see p. 11), in order to efficiently and effectively prepare for the audit and properly train the audit team members, the audit team leader should create and prepare a series of audit tools and templates. Some example tools and templates are provided in the following text. These tool templates are included on CD-ROM, which accompanies this guide, for use and modification as needed. It should be noted that these are sample tools and templates and should be used as a starting point for developing your own project management and training tools. It must also be emphasized that efforts expended during the audit preparation phase will insure the effectiveness, efficiency, and therefore, overall quality of the audit.

Establishing a CGMP Laboratory Audit System. By David M. Bliesner
Copyright © 2006 John Wiley & Sons, Inc.

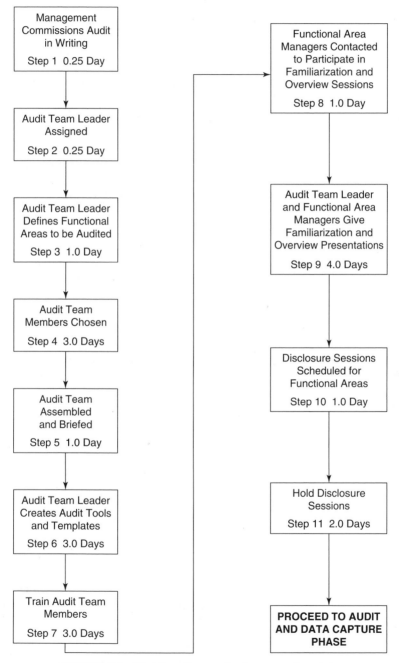

PREPARATION PHASE

Management Commissions Audit in Writing
Step 1 0.25 Day

Audit Team Leader Assigned
Step 2 0.25 Day

Audit Team Leader Defines Functional Areas to be Audited
Step 3 1.0 Day

Audit Team Members Chosen
Step 4 3.0 Days

Audit Team Assembled and Briefed
Step 5 1.0 Day

Audit Team Leader Creates Audit Tools and Templates
Step 6 3.0 Days

Train Audit Team Members
Step 7 3.0 Days

Functional Area Managers Contacted to Participate in Familiarization and Overview Sessions
Step 8 1.0 Day

Audit Team Leader and Functional Area Managers Give Familiarization and Overview Presentations
Step 9 4.0 Days

Disclosure Sessions Scheduled for Functional Areas
Step 10 1.0 Day

Hold Disclosure Sessions
Step 11 2.0 Days

PROCEED TO AUDIT AND DATA CAPTURE PHASE

FIGURE 2.1 Workflow diagram for the preparation phase.

TABLE 2.1 Explanation of Preparation Phase Workflow Diagram Steps

Step	Description	Estimated Duration	Explanation
1	Management commissions audit in writing	0.25 day	The success of any audit depends on management commitment and involvement. In addition, the FDA is very clear in its expectations of management commitment with respect to compliance with CGMPs. Therefore, it is important that management, at some senior level within the organization, formally commissions the audit in writing.
			This commissioning document should include the following sections: (1) Purpose, (2) Start date, (3) End date, (4) Expected deliverables, (5) Designation of audit team leader, and (6) Definition of the team leader responsibilities, level of authority, and accountabilities. The commissioning document should be signed and formally issued to the audit team leader once that individual is selected. Moreover, copies of the document should be circulated to all impacted personnel within the organization. The audit should be a well-publicized event.
			The allocation of one-quarter day to complete the task is based on typical times required to generate an inter-office memorandum. (*Note*: Throughout this guide, the minimum amount of time allocated to any particular tasks is one-quarter day.)
2	Audit team leader assigned	0.25 day	As implied in Step 1, assignment of the audit team leader is part of the audit commissioning process and is performed by senior management. Selection of an audit team leader is critical. The audit team leader is accountable held for the successful completion of the audit within the expected time frame. An individual with good project management and organizational skills is required. The audit team leader need not possess an in-depth understanding of the organization and its operations but should have a good command of laboratory CGMPs and an understanding of the laboratory control system. Previous audit experience is also a plus. Although quality assurance (QA) personnel are often

TABLE 2.1   *(Continued)*

Step	Description	Estimated Duration	Explanation
			considered for such roles, laboratory managers and supervisors should be considered as well.
3	Audit team leader defines functional areas to be audited	1.0 day	The audit team leader works with senior management and department managers to identify those areas that need to be audited. For example, finished product testing laboratories, raw material testing laboratories, product stability testing laboratories, and method transfer laboratories should all be considered for auditing. In addition, any laboratories that may be involved in in-process testing should be included in the audit. Particular attention should be paid to those areas that have known, or are suspected to have, CGMP deficiencies. The selection of the different laboratories to be audited should be communicated to the entire organization in writing by senior management.
			The allocation of 1 day to complete the task may be insufficient for larger facilities with a large number of testing laboratories. Adjust the estimated days required as appropriate.
4	Audit team members chosen	3.0 days	As discussed in Chapter 1, the laboratory control system consists of seven different subelements, namely: (1) laboratory managerial and administrative systems, (2) laboratory documentation practices and standard operating procedures, (3) laboratory equipment qualification and calibration, (4) laboratory facilities, (5) methods validation and technology transfer, (6) laboratory computer systems, and (7) laboratory investigations. Therefore, each of these subelements (as appropriate) needs to be included in the audit.
			Since each of these subelements needs to be included, the ideal composition of the audit team should vary depending upon the subelement and/or the laboratory being

(Continued)

TABLE 2.1   (*Continued*)

Step	Description	Estimated Duration	Explanation
			audited. For example, when the Laboratory Computer Systems subelement is audited, the ideal composition of the audit team would be: (1) A laboratory computer system subject matter expert (SME), (2) a representative from the quality assurance unit, and (3) an outside member (e.g., a consultant or someone from outside the laboratory being audited). However, for smaller organizations with limited resources, forming seven different teams may not be practical. Therefore, at the minimum, the team should include an SME and a representative from QA. The SME should function as the subelement leader who will receive direction from the audit team leader, during the subelement audit as necessary. Audit teams should not have fewer than two people, thus providing a data recorder and an interviewer. This minimum team number requirement insures data are appropriately captured and that the audit progresses in a timely fashion.
			Since execution of CGMP laboratory audits can serve as excellent learning vehicles, consider involving as many personnel as possible. By participating in an audit, one often gains a much better understanding of CGMPs and the structuring and functions of the overall organization.
5	Audit team assembled and briefed	1.0 day	All personnel serving as audit team members should be assembled and briefed as to their individual responsibilities and the responsibilities of all involved parties. This briefing is conducted by the audit team leader. During this briefing the following topics should be covered: (1) Introductions of team members, (2) Review of the commissioning document, (3) Scheduling for training dates, and (4) Scheduled audit start date.
			The allocation of one day to complete the task is given so that the audit team leader has sufficient time to prepare for the in briefing session. The actual session should only take about 1 hour.

TABLE 2.1   (*Continued*)

Step	Description	Estimated Duration	Explanation
6	Audit team leader creates audit tools and templates	3.0 days	In order to efficiently and effectively prepare for the audit and train the audit team members, the audit team leader should create and prepare a series of audit tools and templates. Some of these tools and templates may include: (1) A detailed audit workflow diagram, (2) A weekly audit routine template, (3) A monthly audit schedule template, (4) An audit participant roles and responsibilities matrix, and (5) A detailed audit team member training agenda.
			These types of tools and templates allow for the most efficient use of managers and audit team members' times. Moreover, once they are developed they can be modified at-will (as appropriate) and promulgated during the audit to all of the participants. In short, they can be used as project management tools. In addition, they can also be used for repeat audits executed during the verification and monitoring phases. Some example tools and templates are shown in Figures 2.2–2.4 (see pp. 14, 15, and 24) and Tables 2.2 and 2.3 (see pp. 17 and 21).
7	Train audit team members	3.0 days	Training of audit team members is critical. The better understanding of the audit process all team members possess, the more successful the audit will be. Training is usually conducted by the audit team leader, but may include QA personnel, consultants, personnel from other departments and divisions who have already been through an audit, or any other individuals who may improve the effectiveness of the training.
			At a minimum, training should include: (1) Review of the goals of the audit, (2) An in-depth review of the audit process, (3) Review of roles and responsibilities, (4) Discussion of the working calendar and audit routine, (5) Instruction on data capture and CGMP deficiency documentation, and

(*Continued*)

TABLE 2.1 (*Continued*)

Step	Description	Estimated Duration	Explanation
			(6) Audit strategy development including team member roles and responsibilities, sampling plans, etc. The working calendar is preliminary at this stage and will be finalized following discussions with the functional area managers.
			The allocation of three days to complete the task may be insufficient for larger facilities with a large number of testing laboratories. Adjust the estimated days required as appropriate.
8	Functional area managers contacted to participate in familiarization and overview sessions	1.0 days	These sessions are designed to introduce functional area managers to the audit process. Moreover, it is an opportunity for the audit team leader to get a general understanding of where and how each functional area manager fits into the organization.
9	Audit team leader and functional area managers give familiarization and overview presentations	4.0 days	The audit team leader gives an audit familiarization and overview presentation to all the functional area managers. The presentation should cover: (1) Review of the commissioning document, (2) Overview of the audit process, (3) Data capture procedures, (3) Procedures for reporting findings, and (4) Overview of the corrective and preventive action process. The managers should also be provided with guidelines and/or templates for disclosure session presentations from the audit team leader. This assists them in preparation and execution of disclosure session presentations, which they will be required to give in the future.
			Following the audit team leaders overview presentation, the functional area managers give a very brief description (high level) of who they are, what they do, and what testing their departments or sections are responsible for executing. Managers should provide organizational charts to the audit team leader at this point. Disclosure sessions are used to provide more detail about each

TABLE 2.1 *(Continued)*

Step	Description	Estimated Duration	Explanation
			manager's area and responsibilities. A more detailed explanation of the format for disclosure sessions is presented in Step 11.
10	Disclosure sessions scheduled for functional areas	1.0 day	The audit team leader works with the managers to arrange times that are convenient for both parties.
11	Hold disclosure sessions	2.0 days	Managers make presentations at the disclosure sessions. Disclosure sessions are an opportunity for functional area managers to not only describe their operations in detail but to identify any known CGMP deficiencies. All audit team members, regardless of the subelement, should be in attendance. Lower level supervisors who work for the manager should not attend. This will allow more open and critical disclosure of deficiencies without supervisors or managers feeling challenged and/or threatened. Emphasis should be placed on open disclosure without retribution. The existence of on-going or recently completed corrective actions should be presented as well. (*Note*: Due to human nature and behavior in the modern work environment, auditors should not expect complete disclosure.) An example of a manager's disclosure session agenda is shown in Figure 2.4 (see p. 24).
			Depending on the size of the organization it may take more than two days to work through all the disclosure sessions.

The example audit training agenda shown in Figure 2.2 suggests some key topics on which audit team members should be trained. Some of the items listed will also be used as project management and communications tools as the audit progresses. Detailed explanations of the agenda items (where appropriate) are also presented in the following text.

1. Goals of the Audit
    a. Preparation for FDA Preapproval Inspection (PAI)
    b. Install a Self-audit Program for the QC Laboratories
    c. Improved Operations by Root Cause Analyses and Corrective Action Implementation
2. Review of the Audit Process
    a. Audit Process Workflow Diagram
    b. Subelement Checklists and Checklist Use
    c. Audit Team Membership, Responsibilities, and Deliverables
    d. Audit Notebooks
    e. Weekly Routine
    f. Monthly Calendar
3. Laboratory Audit Form (LAF) Generation Process
    a. Example LAF
    b. LAF Numbering System
4. Subelement Audit Strategy Development
    a. Audit and Interview Schedules
    b. Statistical Sampling and Sampling Plans
5. Miscellaneous

FIGURE 2.2   Example of audit team training agenda.

2.2.1 Goals of the Audit

The goals listed in the example agenda are typical reasons for conducting an audit. However, the goals of an audit need not be limited to the items listed here. As suggested previously, an audit is an excellent training vehicle. This is especially true for quality assurance personnel with no previous laboratory experience. In addition, new managers or supervisors may use an audit as part of their initial familiarization with their new organization and job responsibilities.

Regardless of the goals, conducting an audit and developing and implementing subsequent corrective actions is good business practice, as well as good compliance practice. This should be emphasized during the training session.

2.2.2 Review of the Audit Process

Audit Process Workflow Diagram   The audit process is best described via a flowchart or process diagram. The example audit workflow diagram shown in Figure 2.3 describes the steps normally associated with conducting a CGMP laboratory audit. Each of the steps shown in this diagram should be discussed during the training session. Modifications should be made as necessary to describe the actual audit process to be executed.

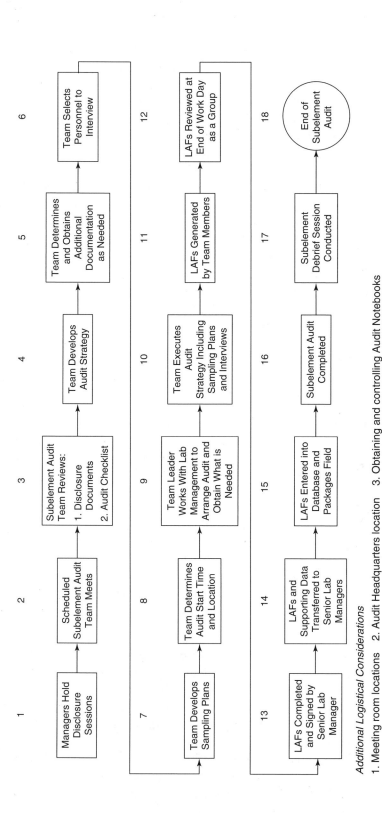

FIGURE 2.3 Example of an audit workflow diagram.

1	2	3	4	5	6
Managers Hold Disclosure Sessions	Scheduled Subelement Audit Team Meets	Subelement Audit Team Reviews: 1. Disclosure Documents 2. Audit Checklist	Team Develops Audit Strategy	Team Determines and Obtains Additional Documentation as Needed	Team Selects Personnel to Interview

7	8	9	10	11	12
Team Develops Sampling Plans	Team Determines Audit Start Time and Location	Team Leader Works With Lab Management to Arrange Audit and Obtain What is Needed	Team Executes Audit Strategy Including Sampling Plans and Interviews	LAFs Generated by Team Members	LAFs Reviewed at End of Work Day as a Group

13	14	15	16	17	18
LAFs Completed and Signed by Senior Lab Manager	LAFs and Supporting Data Transferred to Senior Lab Managers	LAFs Entered into Database and Packages Field	Subelement Audit Completed	Subelement Debrief Session Conducted	End of Subelement Audit

Additional Logistical Considerations

1. Meeting room locations   2. Audit Headquarters location   3. Obtaining and controlling Audit Notebooks
4. Obtaining blank LAFs   5. Copying and control of supporting data   6. Filing of completed LAFs and supporting data

Checklist Use   In addition to using a process diagram, an audit is most efficient and effective when checklists are used. *Checklists should be the core instruments for execution of any audit.* Checklists may be used in several fashions. First, they may be used as a guide to help structure interviews and documentation collection and review. After the interviews and document review, checklists can be completed to determine whether all of the appropriate issues have been covered. Second, they may also be used strictly as checklists, in a question-and-answer format by the auditor, determining whether the laboratory is in compliance with the CGMP components of the subelement. Third, the auditors may hand out checklists to interviewees for completion on their own, with confirmation of the answers during follow-up interviews.

Example checklists corresponding to all seven laboratory control system subelements are included in Appendix I. The audit team leader should review the corresponding checklist with each subelement audit team member as part of the training. Checklists should be modified including, adding, subtracting, or modifying questions as appropriate.

Audit Team Membership, Responsibilities, and Deliverables   Team membership, responsibilities, and deliverables for each individual should be covered during the training session. Table 2.2 shows some of the audit participant roles and responsibilities. The participant roles and responsibilities can be tailored to meet the needs of the particular audit and resources available.

Audit Notebooks   During the course of the audit, it is important to capture findings and observations related to the state of a laboratory's compliance with CGMPs. Audit notebooks serve this purpose and collect raw data during the audit. Notebooks should be issued to all individuals who are auditing and should be the same used as any CGMP scientific notebook. For those individuals who are not familiar with CGMP notebook use, a short introduction or refresher course should be given by the audit team leader ensuring proper capture of all pertinent observations and findings. The more detail recorded by the auditors, the better. This record is used later to help complete laboratory audit forms.

Weekly Routine and Monthly Calendar   Unless carefully planned and managed, a CGMP laboratory audit can take on a life of its own and either consume an inordinate amount of time or languish and never be completed. Therefore, it is important to implement a standard weekly routine and a monthly calendar. Tables 2.3 and 2.4 show an example weekly audit routine and a monthly audit schedule (if the audit continues more than four weeks) that may be used to help manage the audit. All audit team members should be aware of the weekly routine and the task schedule at least a month in advance. These templates can be modified at will and used as communication tools to inform audit team members, managers, and senior management the status of the audit.

TABLE 2.2 Example of an Audit Participant Roles and Responsibilities Matrix

Audit Team Position	Responsibilities	Deliverables	Individual(s) Assigned
Audit team leader	1. Provide logistical support for audit 2. Develop schedules and make role and responsibility assignments 3. Assist managers in understanding audit process 4. Assist managers in preparation of disclosure sessions 5. Supervise audit preparation and execution 6. Collect, collate, and compile LAFs 7. Generate database 8. Create draft audit summary report 9. Communicate status of audit to participants and senior management 10. Complete audit summary report	1. Audit team member roles and responsibility assignment 2. Trained audit team members 3. Written audit work-flow routines and schedules 4. Weekly communication 5. Audit supplies list 6. Signed LAFs 7. Database with LAF entries 8. Draft audit summary report 9. Final audit summary report	J. Casey

(Continued)

TABLE 2.2 *(Continued)*

Audit Team Position	Responsibilities	Deliverables	Individual(s) Assigned
Functional area managers	1. Prepare disclosure session presentations 2. Present disclosure session presentations 3. Educate personnel on lab audit purpose, process, and procedures 4. Provide access to data, personnel, and so on	1. Disclosure session presentations 2. Disclosed deficiencies list 3. Supporting data for disclosure sessions 4. Completed or pending corrective action plans	I. Smooter H. Recardo S. Catts M. Sepuz J. Santis
Audit team subelement subject matter experts (SMEs for subelement)	1. Supervise subelement team 2. Assist in generation of LAFs 3. Insure proper documentation in audit notebooks 4. Monitor collection of supporting data 5. Insure access to audit areas, interview personnel, and supporting data by interacting with lab personnel, supervisors, and managers 6. Participate in daily meetings	1. Documented notebook findings 2. Reviewed draft LAFs 3. Supporting data for LAFs	Laboratory managerial and administrative systems (**MS**) J. Santis Laboratory documentation practices and standard operating procedures (**OP**) A. Qucey Laboratory equipment qualification and calibration (**LE**) J. Smyth Laboratory facilities (**LF**) U. Smith

Role	Responsibilities	Outputs	Members
	7. Insure adherence to schedules		Methods validation and technology transfer (**MV**) E. Vazquez Laboratory computer systems (**LC**) E. Smith Laboratory investigations (**LI**) R. Mazey
Quality assurance audit team member	1. Assist with audit 2. Generate LAFs and collect supporting data 3. Participate in daily meetings and debriefs	1. Documented findings 2. Draft LAFs	A. Lavio
Outside member audit team member (consultant or external department representative)	4. Assist with audit 5. Generate LAFs and collect supporting data 6. Participate in daily meetings and debriefs 7. Provide expertise on specific Subelements 8. Provide perspective	1. Documented findings 2. Draft LAFs	R. Gillen (**LC**) T. Johnson (**LF**) N. Ran (**LE**) V. Dooby (**MV**) J. Alza (**LI**) D. Blistex (**OP, MS**) Additional members as needed

2.2.3 Laboratory Audit Form (LAF) Generation Process

The laboratory audit form (LAF) is the formal data capture instrument. LAFs are generated at the end of the audit day (if at all possible) using notes (raw data) captured in the audit notebook. Details and examples of LAFs are discussed in Chapter 3.

2.2.4 Subelement Audit Strategy Development

Audit and Interview Schedules Each of the seven subelement audits teams are responsible for negotiating audit schedules, scheduling their audits, arranging for interviews, and completing the audit in a timely fashion. Some basic interviewing techniques should be taught to the team members by the audit team leader.

Statistical Sampling and Sampling Plans Statistical sampling can be a powerful tool for use in selecting documents, data, choosing interview candidates, and so on. The use of statistical sampling and development of sampling plans is covered in Chapter 3. The manager's disclosure session can set the tone for an audit, and therefore serves as a critical function.

The example of a manager's disclosure session agenda shown in Figure 2.4 suggests some key topics, which should be discussed by the functional are managers during the disclosure sessions. Detailed explanations of the agenda items are listed in the following text.

1. *Introduction of manager and the major operational aspects of their functional area.* The functional area manager should be able to give a cogent, clear, and brief explanation of the major operational aspects of their functional areas.
2. *Description of functional area's placement in the organizational structure.*

   Location in Company High Level Organizational Chart In addition to not being able to summarize the major operational aspects of their operations, managers who cannot reproduce or clearly describe their organizational structure are those managers who are not in control of their operations. CGMP is equal to showing you are in control.

   Section Names and Functions Managers should be able to identify work sections and their functions and outputs.

   Subsection Names and Functions These are the lowest identifiable operation units within the organization. Managers should be able to define and identify these subsections, name them, and define their functions and outputs.

TABLE 2.3 Example of a Weekly Audit Routine

Monday	Tuesday	Wednesday	Thursday	Friday
15:00–17:00: Subelement audit team(s) meet and:	**08:00–09:00**: Subelement audit team(s) meet with appropriate lab managers and discuss the following:	**08:00–16:00**: Subelement audit begins:	**08:00–16:00**: Subelement audit begins:	**08:00–10:00**: Audit debrief performed by Subelement team leader(s). Discussions include:
1. Review documents provided in disclosure session	1. Clarification of any documents/information provided during disclosure	1. Team members continue executing sampling and audit plans	1. Team members complete sampling and audit plans	1. Summary of LAFs to include:
2. Review assessment checklist	2. Obtain additional documents as needed	2. Continue documenting in audit notebooks and collect supporting data	2. Positive and negative findings are documented in audit notebooks	a. Immediate action requirements
3. Develop subelement audit strategy	3. Define audit schedule for the week	3. Subelement team leaders continue drafting (handwritten) laboratory audit finding forms (LAFs)	3. Sufficient supporting data collected, copied, and collated to support negative and positive findings	b. LAFs that are linked and grouped and can be corrected in unison
4. Determine additional document needs	**09:00–16:00**: Subelement audit begins:		4. Draft (handwritten) LAFs completed by team members	c. Potential corrective action/path forward
5. Determine personnel to be interviewed	1. Team members execute sampling and audit plans	**16:00–17:00**: All Subelement audit team(s) meet to discuss:	**16:00–17:00**: All Subelement audit team(s) meet to discuss:	2. General observations and trends
6. Develop sampling plan(s)	2. Positive and negative findings are documented in audit notebooks	1. The day's operations	1. The day's operations	3. Additional audit needs for particular areas
7. Discuss how to execute sampling plan(s)	3. Sufficient supporting data collected, copied, and collated to support negative and positive findings	2. Draft LAFs	2. Draft LAFs	
8. Time and place to start audit	4. Draft (Handwritten) laboratory audit finding forms (LAFs) completed by Subelement team members	3. Potential crossover audit findings	3. Potential crossover audit findings	
9. Determine need for partners, group audits, and so on		**17:00–18:00**: Subelement team leader meets with lab manager to discuss:	**17:00–18:00**: Subelement team leader meets with lab manager to discuss:	
		1. Provide handwritten copies of LAFs	1. Provide handwritten copies of LAFs	
		2. Discuss immediate action findings		

(Continued)

TABLE 2.3 *(Continued)*

Monday	Tuesday	Wednesday	Thursday	Friday
	16:00–17:00: All Subelement audit team(s) meet to discuss: 1. The day's operations 2. Draft LAFs 3. Potential crossover audit findings **17:00–18:00:** Subelement team leader meets with lab manager to discuss: 1. Provide handwritten copies of LAFs 2. Discuss immediate action findings		2. Discuss immediate action findings	

TABLE 2.4 Example of a Monthly Audit Schedule

Sunday	Monday	Tuesday	Wednesday	Thursday	Friday	Saturday
						1
2	3 **15:00–17:00:** Audit preparation disclosure session by lab managers	4 Computer systems audit / Laboratory facilities audit	5 Computer systems audit / Laboratory facilities audit	6 Computer systems audit / Laboratory facilities audit	7 Computer systems debrief / Laboratory facilities debrief	8
9	10 **16:00–17:00:** Audit preparation meeting	11 Laboratory equipment / Operating procedures	12 Laboratory equipment / Operating procedures	13 Laboratory equipment / Operating procedures	14 Laboratory equipment debrief / Operating procedures debrief	15
16	17 **16:00–17:00:** Audit preparation meeting	18 Laboratory management systems	19 Laboratory management systems	20 Laboratory management systems	21 Laboratory management systems audit debrief	22
23	24 **16:00–17:00:** Audit preparation meeting	25 Management systems audit / Laboratory investigations	26 Management systems audit / Laboratory investigations	27 Management systems audit / Laboratory investigations	28 Management systems audit debrief / Laboratory investigations audit debrief	

1. Introduction of Manager and The Major Operational Aspects of Their Functional Area
2. Description of Functional Area's Placement in The Organizational Structure
   a. Location in Company High-level Organizational Chart
   b. Section Names and Functions
   c. Subsection Names and Functions
3. Summary of Personnel and Skills
   a. Name
   b. Title
   c. Primary Job Function and Responsibilities
   d. Education and Experience Summary
   e. For All Personnel in Manager's Area(s)
4. Description of Work and Workflow
   a. What You Do (All Areas)
   b. Who Performs the Tasks (Each Task)
   c. Output/Finished Product (Data, Reports, etc.)
   d. Who are Your Customers (Who Receives the Output/Finished Product)
   e. Collaborations and extra-site relationships
5. Disclosure of Known Deficiencies
   a. Direct CGMP
   b. Non-CGMP Which May Lead to CGMP Violations
6. Interim Corrective Actions
   a. Currently in Place
   b. Planned
7. Supporting Documentation
   a. Provide Suggested Examples
   b. Other Items as You See Necessary
   c. Examples of Supporting Documentation

   Personnel
   Resumes
   Training Records
   Training Curricula
   Organization Chart
   Job Descriptions
   Offer/Acceptance Letters

   Procedures
   SOPs
   Training Records
   Training Curricula
   Guidance documents
   Flow Diagrams
   Level II and Level III documents

FIGURE 2.4   Example of manager's disclosure session agenda.

Equipment/Instruments

Calibration Records

Maintenance Schedules and Records

Logbooks (e.g., use, cleaning)

Validation Documents/IQ, OQ, PQ

SOPs

Methods

Analytical Methods Validation Protocols and Reports

Tech Transfer Protocols and Reports

Compendial Methods Qualification

Sampling Plans

Test Procedures

Specifications

Out-of-Specification (OOSs) Investigation Reports

Reagents, Reference Standards, and Supplies

Specifications

Qualification Schedules and Protocols

Testing Methods

Certificates of Analysis

Supplier Qualifications

Supplier Audit Schedules and Reports

Facilities

Drawings

Validation Documents

Water System Data

Computer System Validation Records

FIGURE 2.4   (*Continued*)

It is often interesting to note the manager's interpretation of the organizational chart and what personnel at lower levels within the organization perceive it to be. Make notes of such differences during the audit, as this could be an indication of both administrative and CGMP deficiencies.

3. *Summary of personnel and skills.* As part of the organizational chart review, the manager should provide a complete listing of the name, title, primary job function, and responsibilities of each report (including direct and indirect reports). The manager should also provide a summary of their education and experience.

4. *Description of work and workflow.* Managers should provide complete descriptions of the tasks performed in all of their functional areas, who

routinely perform the tasks, and the output/finished product (data, reports, etc.) at completion of execution of the tasks. As with the organizational structure, the manager's interpretation of work and workflow and what personnel at lower levels within the organization perceive it to be may be different. Audit team members should make notes of such differences during the audit because this could be an indication of both administrative and CGMP deficiencies.

Who Your Customers Are   Managers should be able to define who receives the output/finished product, which results from the efforts of all their personnel. This may not be an easy task for many managers because several customers may exist for any given output.

Outsourcing, Collaborations and Extra-Site Relationships   Due to the global nature of most companies, many laboratories are involved in outsourcing, collaborative, and extra-site relationships. Insure that the managers explain these during their disclosure sessions. Many times lack of management of these relationships leads to lack of CGMPs compliance.

5. *Disclosure of known deficiencies.*

Direct CGMP   Some deficiencies will be direct violations for 21 CFR Parts 210 and 211; for example, failing to have written procedures for specific tasks. These findings are straight forward and easily identified by the managers.

Non-CGMP Which May Lead to CGMP Violations   Some deficiencies may not be direct violations of the Current Good Manufacturing Practices Regulations, but if left unaddressed, could lead to such violations. For example, the fact that laboratory personnel are working 6 and 7 days a week is not a direct violation of CGMPs. However, this may be an indication that the laboratory does not have sufficient personnel resources to execute the necessary tasks, which is a violation of the regulations. These types of deficiencies need to be addressed as well.

6. *Interim corrective actions.* Any planned or existing corrective action plans should be identified by the manager. Any success stories for implementation and completion of past corrective actions should be highlighted as well.

7. *Supporting documentation.* Managers are encouraged to provide as much supporting documentation for their disclosed findings as well as general information that may assist the auditors during execution of the audit. The list shown in Figure 2.2 has some suggestions for possible supporting documentation.

CHAPTER 3

AUDITING AND DATA CAPTURE

3.1 ADDITIONAL AUDIT PREPARATION

Chapter 2 was dedicated to the steps needed to prepare for the audit including organizing the audit team, defining work functions, assigning roles and responsibilities, conducting audit familiarization and overview sessions, and performing audit team training. An important part of the audit team training should include instruction on data capture and CGMP deficiency documentation. In addition, time should be spent on development of sampling strategies and plans since many laboratory personnel are not familiar with the technique.

Prior to describing in detail the audit and data capture phase, these two topics need to be addressed more fully. The following sections describe the process of identifying deficiencies, how they are initially documented, how they are formally documented, and how they are ultimately used to improve compliance. These sections also describe the use of random statistical sampling to improve the efficiency and overall quality of the audit.

3.1.1 Data Capture and CGMP Deficiency Documentation

As mentioned in Chapter 2, during the course of the audit it is important to completely and properly capture findings and observations related to the state of a

Establishing a CGMP Laboratory Audit System. By David M. Bliesner
Copyright © 2006 John Wiley & Sons, Inc.

laboratory's potential lack of compliance with CGMPs. Deficiencies or gaps may be divided into two major categories, namely *Critical* and *Noncritical*. Critical gaps are findings which would most likely result in a Form 483 Observation by FDA. Noncritical gaps have to do with administrative systems and practices which can ultimately lead to degraded compliance with CGMPs.

During the audit, gaps are identified in several fashions. First, they may be discovered during completion of the subelement checklists (see Appendix I). Second, they may be revealed throughout the course of the audit during document review and interviews. Third, they may be disclosed by the managers during the disclosure sessions. Fourth, they may have been previously documented (e.g., from previous internal or external audits or findings by the FDA). The mechanics of documenting gaps discovered via each of these mechanisms is as follows:

Subelement Checklists   In this case, deficiencies are indicated directly on the subelement checklist by placing a check in the appropriate box. If additional detail is needed, it can be written on the checklist or documented and referenced in the audit notebook. Supporting data should be attached to the checklist as appropriate. References to the supporting data can be made on the checklist or in the audit notebook. It should be emphasized again that, whenever possible, the audit checklists should be the core instruments for execution of the audit. A basic tenant of quality auditing, regardless of the industry or regulatory agency under which the scrutiny of the organization's operations fall, is to use checklists during execution of the audit.

Document Review and Interviews   Deficiencies identified in this fashion should be documented in the audit notebook. Supporting data should be referenced in the notebook, labeled, and filed separately if it cannot be affixed to the notebook itself.

Manager Disclosure Sessions   Deficiencies identified by the managers during disclosure sessions should be supported by data and confirmed during the course of the audit. This confirmation is then recorded in the audit notebook and the supporting data filed as appropriate.

Previous Audit Results and FDA Observations   As with deficiencies identified by the managers during the disclosure sessions, previous audit findings or FDA observations should be supported by data and confirmed during the course of the audit. This confirmation is then recorded in the audit notebook and the supporting data filed as appropriate.

As described previously, in many cases, the audit notebook is used as the primary data-capture instrument. Audit notebooks serve as the point of raw data collection during the audit. The types of notebooks used are a matter of personal preference; however, the use of bound books with permanent page numbers is recommended. Record books and laboratory notebooks typically used for documenting instrument usage, maintenance and calibration, or recording of laboratory data, solution preparations, and so on are also recommended. These notebooks should be issued and treated as controlled documents. Auditors should be encouraged to use the notebook to record any and all observations made during the audit. There is no such thing as over documentation. All observations should be made in pen and corrections made as per standard CGMPs. Again, supporting documentation can be affixed in the notebook, if appropriate.

Once a gap is identified and captured, it will then be formally documented on a laboratory audit form (LAF) or similar data capture form. The LAF is the formal data capture instrument used during the audit. The design of the LAF can be tailored to meet the needs of the organization and the audit. Figure 3.1 shows a completed example LAF.

A detailed explanation of the forms contents follows.

1. *Laboratory Audit Form (LAF) Title of Form.*

    Confidential: The LAF header should include a statement as to the confidential nature of the information contained therein.

    Laboratory Control System Subelement: Laboratory Managerial and Administrative Systems Step 1.1.1. This section of the header links the subelement checklist to the finding or gap. Specifically, in this example, if you go to the subelement 1, Laboratory Managerial and Administrative Systems checklist and look for Step 1.1.1, you will find the question: "Are current organization charts available and accurate?"

    LAF No.: LCS-MS-001   This is the LAF number which is used to track the gap from its initial observation to completion of the corrective action. This number is a key entry into the corrective action project plan (CAPP) database. (*Note*: the CAPP is discussed in Chapter 4.) The LAF numbering system is as follows: LCS denotes Laboratory Control System. This is included in the name in the event the laboratory audit is part of a larger quality system audit, which may include other systems such as the Production System, Packaging and Labeling System, and so on. MS denotes the Laboratory Managerial and Administrative Systems subelement of the Laboratory Control System. -001 denotes the first finding for the Laboratory Managerial and Administrative subelement. The remainder of the LAFs will be in numerical order. It is possible to design a numbering system that links directly to the audit subelement checklists. However, this makes it difficult to assign numbers to findings that are not the result of a checklist item deficiency.

Laboratory Audit Form (LAF)

CONFIDENTIAL: This document is not to be distributed or copied except with the written permission (Your Company Site Management)

Laboratory Control System Subelement:	Laboratory Managerial and Administrative Systems Step 1.1.1

		LAF No.:	LCS-MS-0001

Audit Notebook No.:	AN-002	**Page No.:**	33 to 37, 41

Auditor(s):	J. Carrie, J. Feliz

Personnel Interviewed:	Jenny Smith, Lab Training Supervisor

Deficiency Identified During Disclosure Session?: ☐ Yes ☒ No

LAB DOCUMENT(S) REVIEWED:

Most recent Laboratory Organization Chart dated 15 May 2000.
SOP RRR-02-001 ADMINISTRATION OF CHANGES TO LABORATORY ORGANIZATION CHART (Canceled)

AUDIT CHECKLIST FINDINGS:

Checklist Item Description: Step 1.1.1 Are current organization charts available and accurate?

Brief Deficiency Description: Organization charts have not been updated since May 2000.

Priority Assignment Based on Impact: ☐ Immediate ☒ High ☐ Routine ☐ Low

Potential Root Cause: Lab Manager does not currently have system in place to perform periodic review of time sensitive documents.

Potential Corrective Action: Work with QA document control specialist to generate a list of time sensitive documents which need periodic review. Schedule review and add to laboratory yearly commitment schedule.

Linkage to Other Quality Systems Elements: 4 – Documentation, 3 – Change Management

Additional Comments or Clarifications: None

FIGURE 3.1 Example of a completed laboratory audit form (LAF).

ADDITIONAL ITEMS:

Brief Deficiency Description: None

Supporting References: na

Priority Assignment Based on Impact: ☐ Immediate ☐ High ☐ Routine ☐ Low

Potential Root Cause: na

Potential Corrective Action: na

Linkage to Other Quality Systems Elements: na

Additional Comments or Clarifications: na

LINKAGES:

Source: Description and Identification:

☐ None

☐ 483 Observations

☐ Previous Audits

☒ Gap Analyses   Finding Step 1.14.9-XX001771

☐ Other LAF Findings

☐ Other Links

ATTACHED SUPPORTING DATA LIST:

Copy of Lab Organization Chart date 15 May 2000

Prepared By: _____   Date: _____

Lab Management Review: _____   Date: _____

QA Review: _____   Date: _____

Database Entry: _____   Date: _____

FIGURE 3.1   (*Continued*)

Audit Notebook No. and Page No.: This is the number of the controlled audit notebook and the page numbers where observations were recorded. Multiple notebook and page numbers may be shown here in that the same observation may be made by several different individuals and/or on more than one occasion.

Auditor(s): Self explanatory. More than one person can be designated auditor.

Personnel Interviewed: List all personnel interviewed during the course of making this particular observation.

Deficiency Identified During Disclosure Session: If a manager disclosed the deficiency, it needs to be confirmed during the audit. This information is documented on an LAF.

2. *Lab Document(s) Reviewed*: All documents reviewed which support the finding should be included here.

3. *Audit Checklist Findings*

Checklist Item Description: The checklist item, as it is written on the checklist, is reproduced here. If there is no correlation to a checklist finding, "Na" is entered and the bottom half of the form is completed under "ADDITIONAL ITEMS".

Brief Deficiency Description: A concise description of the deficiency is given in this space. As many details as necessary are included to insure complete documentation.

Priority Assignment: ☐ Immediate ☒ High ☐ Routine ☐ Low The priority assignment indicates the severity of the gap. *Immediate* implies a significant violation of the CGMPs that could significantly impact the safety and efficacy of the finished product. *High* implies those gaps that could result in an FDA 483 finding or worse. *Routine* and *Low* are gaps associated with general administrative systems and practices, which by themselves, are not violations of the CGMPs but if not addressed can lead to failures.

Potential Root Cause: The root cause is the fundamental difficulty that leads to the deficiency. For example, repeated out-of-specification (OOS) investigations may indicate that a method is not operating as per its intended use and thus is not properly validated. Every gap should have at least one root cause. Although it might not be possible to identify the root cause(s) during the audit, some effort should be made to assign one at this point. Root cause determination is often not straightforward, but the auditor can work with personnel in the impacted areas to make a reasonable first assessment of the root cause.

Potential Corrective Action: This is the auditor's recommendation on how the gap may be remediated based on the information at-hand. As with the root cause determination, the auditor can work with personnel in the impacted areas to make a reasonable first offering as to the potential corrective action.

Linkage to Other Quality System Elements: In many instances, the laboratory control system and its subelements are intertwined with the other quality system elements. If there is linkage, then this linkage should be stated. For example, difficulties associated with outdated methods, are intertwined with overall quality system.

Additional Comments or Clarifications: Add as appropriate.

4. *Additional Items*: This section is for findings that are not correlated directly to the subelement checklists. Each section is completed in a similar fashion to the sections for those items which do correlate to a checklist. If checklist correlation exists, simply enter "Na" as appropriate.

5. *Linkages*: This section links the current finding to other potential sources. This is entered into the database and insures that comprehensive systems based solution is implemented that covers all previously documented deficiencies as well as the current finding. The appropriate blocks are checked as necessary and supporting explanation or document of the linkage is included.

6. *Attached Supporting Data List*: The LAF should be the cover page of a data package. This package should include supporting data identified during the audit and documentation of the deficiency. As the audit is completed, and the corrective and preventive actions (CAPAs) forms are completed, the corrective action information and documentation should be included in the data package as well. Upon completion, the entire life cycle of the gap is captured in this data package.

7. *Footer Information*

   Prepared By: This is either the auditor or a support person who actually generated the LAF.

   Lab Management Review: This is the signature of the laboratory manager whose area in which the gap was identified.

   QA Review: A member of the quality assurance department should review and sign all LAFs. This is true even if QA was not involved in the actual audit.

   Database Entry: This is the signature and date of the person who entered the finding into the corrective action project plan (CAPP) database for tracking.

As mentioned above, laboratory audit forms are the formal data capture instrument for the audit. Because of this, LAFs are ultimately used to improve the laboratories level of compliance with CGMPs. This is accomplished by converting the LAFs to an audit summary report (ASR). Moreover, certain critical information included in the LAFs is also entered into a corrective action project plan (CAPP) database which is in turn used to generate a corrective action project plan itself, which is ultimately used to implement corrective actions and preventive actions (CAPAs). Details concerning the format, content, and creation of the ASR are described in Chapter 4. Creations of the CAPP and subsequent CAPAs are delineated in later chapters of this guide. Regardless, the LAF-to-corrective action process is shown in Figure 3.2.

As mentioned at the beginning of this chapter, prior to describing in detail the Audit and Data Capture phase, the use of statistical sampling to improve the efficiency and overall quality of the audit also needs to be considered. The following section describes the concept of random statistical sample, its utility in conducting and audit, and how one can approach the audit using these tools.

3.1.2   Use of Random Statistical Sampling to Improve the Efficiency and Overall Audit Quality

Random statistical sampling is the application of a random number generator in combination with a statistically determined sample set size, to choose what data, personnel, documents, and so on are to be scrutinized during the audit. The use of random statistical sampling during execution of the audit, serves several purposes, namely:

1. It forces the auditor to carefully choose what data, documents, personnel, and so on are to be sampled and subsequently reviewed or interviewed. In other words it imposes structure on the audit approach. This can be very important in that it is often difficult to determine where to start when conducting an audit.
2. The random selection of data, documents, personnel, and so on imposed on the chosen statistical population gives an unbiased or true picture of the entire state of compliance (whatever aspect that may be) of the entire chosen population. Biases on the part of the auditor or personnel supplying the data are minimized.
3. The random selection of data serves as a means to depersonalize an audit. During the execution of an audit, sometimes resistance exists on the part of the persons, sections, or departments being audited to become fully engaged in the audit process. Due to politics and previous

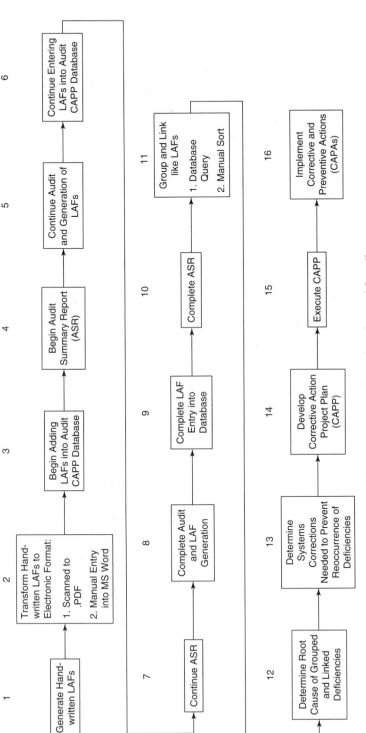

FIGURE 3.2 LAF-to-corrective action plan workflow diagram.

1. Generate Hand-written LAFs

2. Transform Hand-written LAFs to Electronic Format:
   1. Scanned to .PDF
   2. Manual Entry into MS Word

3. Begin Adding LAFs into Audit CAPP Database

4. Begin Audit Summary Report (ASR)

5. Continue Audit and Generation of LAFs

6. Continue Entering LAFs into Audit CAPP Database

7. Continue ASR

8. Complete Audit and LAF Generation

9. Complete LAF Entry into Database

10. Complete ASR

11. Group and Link like LAFs
    1. Database Query
    2. Manual Sort

12. Determine Root Cause of Grouped and Linked Deficiencies

13. Determine Systems Corrections Needed to Prevent Reoccurrence of Deficiencies

14. Develop Corrective Action Project Plan (CAPP)

15. Execute CAPP

16. Implement Corrective and Preventive Actions (CAPAs)

interactions, a sense that their work is being singled out and therefore the subject of unfair scrutiny, may exist. Conversely, auditors are sometimes threatened with the prospect of reviewing certain data or interviewing personnel of higher position within the organization. Random statistically sampling has a tendency to eliminate any perception of unfairness or reluctance on the part of the auditor because it's random. You can always blame the math instead of the auditor for the data, documents, and personnel chosen in the review.

The best way to explain these three advantages is to look at an example application of the technique.

An Example Sampling Plan   As an example, let us assume an auditor desires to review a laboratory's level of compliance with subelement 2, — laboratory documentation practices and standard operating procedures (OP). Specifically, he or she wish to look at the quality of notebook documentation across all product release testing laboratories.

Tackling this review in a systematic fashion is no small feat, particularly when a large number of laboratories are involved. However, by using random statistical sampling, the task can become much more manageable. To solve this problem, the auditor may take the following steps (please refer to Figure 3.3 while following the steps).

Step 1 Identify Labs for Which the Manager is Responsible   In this example, the auditor has identified four separate release testing laboratories that fall under the auspices of the laboratory manager.

Step 2 Identify Each Product Associated With Each Lab   Each of the identified laboratories may be responsible for release testing of different or the same product. The auditor needs only to identify the products being tested by each separate lab without concern for duplication of product or testing. Remember, the exercise is to obtain a random selection of notebooks across all laboratories.

Step 3 Identify Each Test Being Conducted for Each Product Within Each Lab   In this example, the auditor has determined that five release tests are performed on product 1 in release lab 1, three tests for product 1in release lab 3, and so on. Again, the auditor needs only to identify the tests being performed on each product for each lab without concern for duplication of selection of tests.

All Product Release Labs

Release Lab 1			Release Lab 2	Release Lab 3		Release Lab 4
Product 1	Product 2	Product 3	Product 1	Product 1	Product 2	Product 1
1 Test 1	**6 Test 1**	**10 Test 1**	13 Test 1	18 Test 1	21 Test 1	22 Test 1
2 Test 2	**7 Test 2**	11 Test 2	14 Test 2	**19 Test 2**		23 Test 2
3 Test 3	**8 Test 3**	12 Test 3	15 Test 3	20 Test 3		24 Test 3
4 Test 4	9 Test 4		16 Test 4			
5 Test 5			17 Test 5			

1. Identify labs for which the manager is responsible.
2. Identify each product associated with each lab.
3. Identify each test being conducted for each product within each lab.
4. Create a numbered list from 1 until the end, for each test for each product for each lab (see diagram).
5. Identify the total number of tests for all products, this is the sample size N.
6. Find the square root of the total number of all tests, N, for all products for all labs.
7. Round the square root of N, up to the nearest whole number.
8. Add 1 to the rounded number: This is the number of tests which will be randomly sampled.

For Example Using the Diagram:

24 total tests exist for all of the products in all of the manager's labs release testing labs. The square root of 24 is 4.899 which rounds to 5. Add 1 and the total to sample is 6. Using a random number generator, such as RAND BETWEEN (1,24) from EXCEL, select 6 choices between 1 and 24, such as: 19,7,10,6,2,8 (in bold).

9. For the 6 selections, identify all the outputs for the last 6 months of testing for those selected products analyzed by those tests. Output may be notebooks, chromatograms, reports, etc.
10. Choose an output, such as notebooks, and identify the total number of notebooks used during the last 6 months for the 6 tests chosen randomly.
11. If the total number of notebooks used is small (e.g., 6 or less) review all notebooks for compliance with CGMP documentation practices.
12. If the total number of notebooks is large, randomly sample the books using the technique demonstrated above.

FIGURE 3.3 Example of a sampling plan development process: Selecting laboratory notebooks for review.

Step 4 Create a Numbered List from 1 Until the End, for Each Test, for Each Product, for Each Lab   Number all of the tests in numerical order, starting with release lab 1, product 1, test 1 and continuing to the end with release lab 4, product 1, test 3. This will give a list of 24 tests. Some may be duplicates, but it does not matter. (Refer to Figure 3.3 for a graphical explanation of the test numbering scheme.)

Step 5 Identify the Total Number of Tests for All Products: the Population, N   In the example, shown in Figure 3.3, the total number of tests (some of which may be duplicates) is 24. N is the population.

Step 6 Find the Square Root of the Total Number of all Tests, N, for All Products for All Labs   In order to perform statistical sampling, the sample size is derived by taking the square root of the population, N, rounding this number to the nearest whole number, and adding 1. In this example, $N = 24$.

Step 7 Round the Square Root of N, Up to the Nearest Whole Number   The square root of $24 = 4.89897$, which rounds to 5.

Step 8 Add 1 to the Rounded Number   This is the number of tests that will be randomly sampled. Add 1 to 5 and get 6, which is the sampling set size used during random sampling.

Step 9 For 6 Selections, Identify all the Outputs for the Last 6 Months of Testing for Those Selected Products Analyzed by Those Tests   In this example we have chosen to look at laboratory notebooks. Output may be notebooks, chromatograms, reports, and so on.

Step 10 Choose an Output, such as Notebooks, and Identify the Total Number of Notebooks Used During the Last 6 Months for 6 Tests Chosen Randomly   One of the ways to randomly choose these notebooks is to use the random number generator function in Microsoft Excel® spreadsheet software. In our example:

Sampling for $N = 24$
$SQR(24) = 4.898979486$, which rounds to 5, plus $1 = 6$.
Enter values into spreadsheet function: $=RANDBETWEEN(1,24)$

Hit F9 key 6 times to populate a sample table such as:

Choice Number	Random Selection	Lab, Product, and Test
1	2	Lab 1, Product 1, Test 2
2	6	Lab 1, Product 2, Test 1
3	7	Lab 1, Product 2, Test 2
4	8	Lab 1, Product 2, Test 3
5	10	Lab 1, Product 3, Test 1
6	19	Lab 3, Product 1, Test 2

This gives you 6 randomly selected choices between 1 and 24. Since you numbered all tests from 1 to 24, the random selections can be correlated to a lab, product, and test. You now identify the laboratory notebooks for each selected lab, product, and test over the last 6 months.

Step 11 If the Total Number of Notebooks Used is Small (e.g., 6 or less) Review All Notebooks for Compliance with CGMP Documentation Practices The auditor will have to use some judgment at this stage and determine what an appropriate review will encompass. When in doubt, obtain more data.

Step 12 If the Total Number of Notebooks is Large (e.g., 100 or larger), Randomly Sample the Books Using the Technique Demonstrated For very large populations (e.g., 1000 or greater) a maximum sampling of 30 choices is usually sufficient.

Remember, sampling is a tool, and therefore, can be used as deemed appropriate by the auditor. Larger or smaller sample sizes can be used if deemed necessary.

3.2 PROCEDURE

Having described data capture and CGMP deficiency documentation, as well as use of random statistical sampling to improve the efficiency and overall quality of the audit, it is now possible to describe the details of the auditing and data capture phase. The steps in this process are shown in Figure 3.4 and described in Table 3.1.

Some details for each step are summarized in Table 3.1. Note that time estimates are not included in this table due to the extremely variable nature of the amount of time which may be required during the audit and data capture phase.

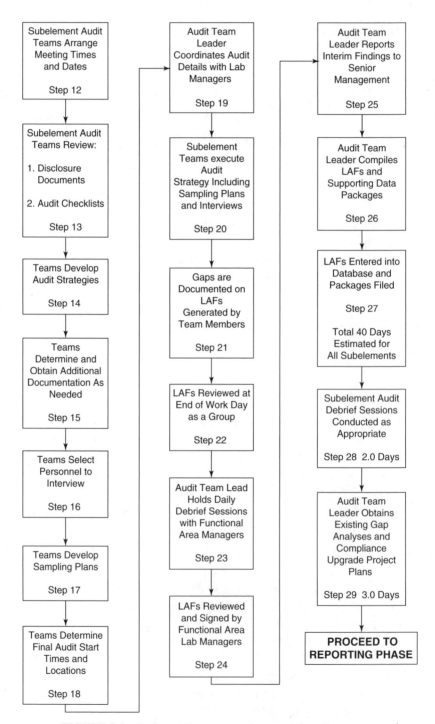

FIGURE 3.4 Audit and data capture phase workflow diagram.

TABLE 3.1   Explanation of Audit and Data Capture Phase Workflow Diagram Steps

Step	Description	Explanation
12	Subelement audit teams arrange meeting times and dates	Recall that the subelement audit team subject matter expert (SME) usually serves as the subelement team leader. The subelement leader with the assistance of the audit team leader finalizes the audit routine and calendar shown in Chapter 2. Actual times and dates are negotiated with functional area supervisors, managers, and the personnel within their organizations.
13	Subelement audit teams review: 1. Disclosure documents 2. Audit checklists	The subelement team members meet to review the disclosure documents provided by the managers. Since the audit should be based on the subelement checklists, a detail review and discussion of the checklist contents ensues. Team members are encouraged to expand the checklists based on their own experiences.
14	Teams develop audit strategies	When possible, the use of organizational charts and the application of random statistical sampling as demonstrated here should be used as a basis for developing audit strategies.
15	Teams determine and obtain additional documentation as needed	During the audit strategy sessions, additional documentation may be needed by the audit team members to better plan for the audit.
16	Teams select personnel to interview	Personnel to interview are usually chosen based upon the subelement topic, organizational chart, and through random sampling.
17	Teams develop sampling plans	As mentioned previously, random statistical sampling should be used whenever possible to help structure and guide the audit.
18	Teams determine final audit start times and locations	Times and locations may change frequently during an audit. Flexibility and communication of changes in the schedule ensure smooth audit flow. Promulgate changes to the audit routine and schedule as often as necessary.

(Continued)

TABLE 3.1 (*Continued*)

Step	Description	Explanation
19	Audit team leader coordinates audit details with lab managers	Although the SMEs are in charge of the subelements, the audit team leader needs to work closely with the managers to address any difficulties which may arise during the audit. In some circumstances significant deficiencies may be identified and thus create tension between the auditors and the functional area personnel. Also, there may be times when auditors consider certain issues to be gaps, whereas the functional area personnel do not. In these situations, the audit team leader and managers must show leadership and resolve any conflicts.
20	Subelement teams executes audit strategy including sampling plans and interviews	Self explanatory.
21	Gaps are documented on LAFs which are generated by team members	Raw data are recorded in the audit notebooks and formal findings are documented on LAFs. Supporting documentation is also collected at the time of the audit. Ideally, LAFs should be created at the time of their deficiency observation, but this is often not possible during the actual audit. Therefore, time should be included in the audit schedule every day to create LAFs. LAFs need to be generated on an ongoing basis. The audit teams must resist the desire to try and write them all up at the end of the audit.
22	LAFs reviewed at end of work day as a group	All subelement audit teams meet at the end of the day to discuss individual findings. Gaps identified by one subelement may lead insight into potential gaps in another. Such meetings also give the group an opportunity to discuss auditing techniques and skills. Theses meetings also give the audit team leader an idea of the overall status of the audit, how it is progressing, and what he or she may need to do to improve the audit workflow process. It will also identify significant areas of noncompliance which may need to be addressed immediately.

TABLE 3.1 *(Continued)*

Step	Description	Explanation
23	Audit team leader holds daily debrief sessions with functional area managers	Functional area managers and supervisors receive a summary of the findings uncovered during the audit activities for that day. Critical CGMP violations will be identified and addressed immediately.
24	LAFs reviewed and signed by functional area lab managers	Once the LAFs have been completed and reviewed by the audit team leader, functional area managers are given the opportunity to review and comment on them as well. Once the audit teams and managers are in agreement as to LAF contents, managers sign them.
25	Audit team leader reports interim findings to senior management	Senior management should be kept abreast of the audits progress as well as presented with a summary of the findings on a daily basis. CGMP critical findings should be presented to senior management level once enough data has been acquired to confirm the gap.
26	Audit team leader compiles LAFs and supporting data packages	Once the LAFs have been signed by the functional area managers, the audit team leader compiles supporting data packages to be filed with the LAFs. Recall, that when the gaps have been corrected the LAF, the supporting data, the corrective action plan, the corrective actions, and the data which demonstrate the successful implementation of the corrective actions constitutes the contents of a single data package or file. This will demonstrate to FDA the complete and successful remediation of the deficiency.
27	LAFs entered into database and packages filed	Depending upon the size of the facility and the scope and breadth of the audit, it may be necessary to create a relational database to manage the development and implementation of corrective action plans and corrective and preventive actions. Having the database allows you to correlate similar gaps identified across several subelements, and therefore develop systems based solutions as opposed to Band-Aid or single observation solutions. Some suggestions for database fields are shown in Figure 3.5.

(Continued)

TABLE 3.1 (*Continued*)

Step	Description	Explanation
28	Subelement audit debrief sessions conducted as appropriate	At the completion of the audit for a given subelement, a debrief session including the subelement team member, and audit team leader, and the functional area managers should be conducted.
29	Audit team leader obtains existing gap analyses and compliance upgrade project plans	At the completion of all subelement audits, the audit team leader obtains any existing gap analyses, planned or ongoing compliance upgrade efforts, or other effort which are aimed at upgrading the state of laboratory CGMP compliance. These are used to create corrective action project plans and subsequent corrective and preventive actions.

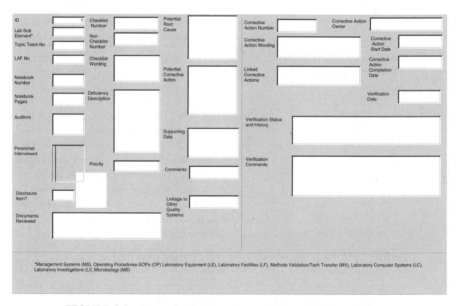

FIGURE 3.5   Example of a laboratory audit form tracking database.

CHAPTER 4

ORGANIZING DATA AND REPORTING THE RESULTS

4.1 PROCEDURE

Once the audit is complete and data have been captured, the process of organizing the findings, presenting them in a coherent summary report, and circulating the report for review begins. The steps in this process are shown in Figure 4.1 and described in Table 4.1.

Although the number of steps in this process are few, the criticality and potential difficulty of generating the summary report cannot be overemphasized. The audit summary report is the final product of the actual audit.

The audit summary report is a very valuable and useful document, not only from a what-it-costs standpoint, but also from a compliance and business efficiency standpoint. Specifically, if the audit is comprehensive and is executed as delineated in the previous chapters, a significant number of labor hours are invested and the direct cost from labor alone can be substantial. However, if performed in a proper fashion, this investment in time and effort has value from a CGMP compliance perspective. Namely, a detailed understanding of your current level of CGMP compliance is provided and be a regulatory agency is shown what you know. From a business standpoint, the audit summary report lays the basis for the most systematic and efficient

Establishing a CGMP Laboratory Audit System. By David M. Bliesner
Copyright © 2006 John Wiley & Sons, Inc.

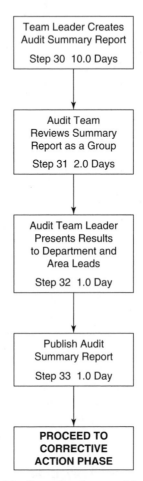

FIGURE 4.1   Reporting phase workflow diagram.

means to upgrade your level of compliance. Not to mention that non compliance in general can be very expensive if it results in significant regulatory action.

It should be emphasized that the process of organizing and reporting the results is a critical phase in that it lays the ground work for developing a future corrective action plan. The greater the effort expended on determining how the data are to be reported, the more effective and straightforward the creation and implementation of the corrective action plan will be.

TABLE 4.1 Explanation of Reporting Phase Workflow Diagram Steps

Step	Description	Estimated Duration	Explanation
30	Audit team leader creates audit summary report	10 days	The first step in the process involves grouping audit findings under the laboratory subelement for which deficiencies were identified. Deficiencies are captured on LAFs during the audit and data capture phase. After the findings have been grouped, the process of creating a report from a standard report template can begin.
			The final format and content of the audit summary report is determined by details and logistics of the audit itself. However, the structure of all audit summary reports should be essentially the same. With this in mind, a template for an example report is shown in Appendix I. Descriptions of the contents of each section of the report are shown in the following sections of this chapter.
			The allocation of 10 days is based on using an existing report format, and full availability of dedicated personnel, and finding reasonable deficiencies. This number can increase significantly if personnel are not dedicated to the audit, or if a significantly large number of findings are documented.
31	Audit team reviews report as a group	2 days	Once the audit team leader has generated the draft audit summary report, they circulate the report to all members of the audit team. Following individual review, the team meets as a group to discuss and agree upon the content of the report. This is done prior to presenting their results to the department area leaders.
			This process should take approximately 2 days, but depends upon the number of findings, the level the complexity of the report, and the criticality of the findings.
32	Audit team leader presents results to department area leaders	1 day	This presentation is done in advance of issuing the audit summary report. The audit team leader and other members of the audit team, as appropriate, present a summary of why and how the audit was conducted, what findings were made, and

(Continued)

TABLE 4.1 *(Continued)*

Step	Description	Estimated Duration	Explanation
			some recommendations of on how the departments are to proceed in correcting deficiencies. The audit team leader needs to take particular care in presenting deficiencies to the group. Due to the potential sensitivity of finding CGMP deficiencies, a certain amount of "push back" should be expected from department management. This meeting sets the tone for implementation of successful corrective action and verification plans.
			This process should take approximately 1 day, but as before, depends upon the number of findings, the level the complexity of the report, and the criticality of the findings. Certain deficiencies maybe determined to be GMP critical and thus impact the safety and efficacy of the products being released to market. These findings will have to be discussed and acted upon immediately and thus may require more than 1 day to address.
33	Publish audit summary report	1 day	Following the presentation of results to the department area leaders, the final report needs is generated, reviewed, and signed by the audit team leader, quality assurance, and management, as appropriate.

4.2 FORMAT AND CONTENT OF THE AUDIT SUMMARY REPORT

As stated in Table 4.1, the final form of the audit summary report (ASR) is determined by the details and logistics of the audit itself. However, the structure of all ASRs should be essentially the same. With this in mind, an example report and a description of the contents of the report are shown in the following sections. Please refer to the audit summary report (ASR) example template shown in Appendix II, while reading the descriptions listed below:

4.2.1 Header

The header should identify your facility name and location in addition to all of the personnel who were involved in the audit. A statement as to the confidential nature of the material included in the report should be made as well.

4.2.2 Background

The background section summarizes the purpose for performing the audit. For example, the audit may be in preparation for an FDA preapproval inspection, (PAI), an upgrade of your existing quality systems, or a continuation of an existing audit program.

4.2.3 Approach

This section should describe all the subelements of the laboratory quality management system reviewed during the audit. As described in previous sections, these elements include:

- Laboratory managerial and administrative systems
- Laboratory documentation practices and standard operating procedures
- Laboratory equipment qualification and calibration
- Laboratory facilities
- Methods validation and technology transfer
- Laboratory computer systems
- Laboratory investigation

It should also discuss the personnel who were involved in the audit, the mechanics of the audit (e.g., use of checklists), and how the findings were documented (e.g., in a notebook with subsequent documentation on an official finding form, such as a LAF).

4.2.4 Report Format

This section discusses how the summaries of each of the subelement findings contained within the body of the report are organized; namely:

- A brief description of the subelement
- An overview of the current practice at the site or each subelement
- A listing of site documents reviewed
- Gaps in the subelement versus checklists or similar quality review documents

TABLE 4.2   Example of a Report Summary Matrix

Checklist Item Number and Description	Gap	In Substantial Compliance?	Potential Root Cause	Potential Corrective Action
3.1.1 Are the laboratories equipped with all of the necessary instruments for the analytical testing to be performed?	Many instruments are awaiting IQ/OQ/PQ and are therefore not available for use.	Na	There is a lack of adequate laboratory resources for performing IQ/OQ/PQ.	Hire or designate personnel to perform IQ/OQ/PQ.
Etc.				

- Additional gaps not correlated to checklists or similar quality review documents
- Potential root causes for the gaps
- Potential corrective action needed to become compliant with CGMPs
- A summary matrix for these steps that can be used to create the corrective action plan

The format of the report can be tailored to fit individual site needs. However, it is strongly suggested that a summary matrix being included for each subelement. The summary matrix is organized as in Table 4.2.

This matrix format greatly enhances the generation of a corrective action plan.

4.2.5 Summary of Results

The summary or results section should capture the total number of findings (checklist gaps and nonchecklist gaps) discovered for all subelements during the audit. As with the individual subelement findings, the summary results should be organized into a matrix as well. Table 4.3 shows what an example format might include.

Nonchecklist gaps are observations made in addition to those findings which were discovered by using an audit checklist or similar review document. In addition to the total findings, a summary of critical versus noncritical gaps should be presented as well.

Any gaps which are designated as critical are those gaps which could potentially warrant a Form 483 observation from an FDA inspector.

TABLE 4.3   **Example of a Results Summary Matrix**

Subelement	No. of Checklist Item Gaps	No. of Nonchecklist Gaps	Total No. of Gaps	Percent of Total Gaps Found
1.0 Laboratory managerial and administrative systems (MS)	20	32	52	17.7
2.0 Laboratory documentation practices and operating procedures (OP)	25	78	103	35.2
Etc.				

TABLE 4.4   **Example of a Summary of Critical versus Noncritical Gaps**

Subelement	Total No. of Gaps	No. of Critical Gaps	No. of Noncritical Gaps
1.0 Laboratory managerial and administrative systems (MS)	52	14	38
2.0 Laboratory documentation practices and operating procedures (OP)	103	14	89
3.0 Laboratory equipment qualification and calibration (LE)	27	10	17
Etc.			

4.2.6   Future Work

The future work section should review the steps required for the implementation of a complete audit, namely:

- Preparation phase
- Audit and data capture phase
- Reporting phase
- Corrective action phase
- Verification phase
- Monitoring phase

Some explanation should be given as a need to continue with the corrective action phase, and the potential resources which may be required to complete the full audit.

4.2.7 Laboratory Controls Subelement Sections

This section then presents the data for each subelement as described in the format section namely:

- Description of subelement
- Current practice
- Site documents reviewed
- Gaps in the system versus audit checklist

The level of detail and breadth of discussion depends upon individual site organizational structure and level of compliance with CGMP.

CHAPTER 5

DEVELOPING AND IMPLEMENTING A CORRECTIVE ACTION PLAN

5.1 PROCEDURE

Once the audit summary report (ASR) is complete, the report is shared with management so that a corrective action project plan (CAPP) and subsequent corrective actions and preventive actions (CAPAs) can be created and implemented. This is the most tedious phase of the audit and requires considerable effort in order to establish an efficient CAPP and implement effective CAPAs. There is a natural tendency at this stage for management to overreact and find individual blame for identified shortcomings. This tendency must be resisted and supplanted by a detailed team-based root cause analysis which leads to clearly defined systems-based corrective actions. This approach leads to long-term sustainable compliance with laboratory CGMPs. The steps in this process are shown Figure 5.1 and described in Table 5.1.

As described in Table 5.1, the process of converting audit findings to systems-based corrective and preventive actions requires the creation of a work breakdown structure (WBS) and translation into an operational project plan. A detailed explanation of the entire process is given through the following example with reference to Figures 5.3, 5.4 and 5.5.

Establishing a CGMP Laboratory Audit System. By David M. Bliesner
Copyright © 2006 John Wiley & Sons, Inc.

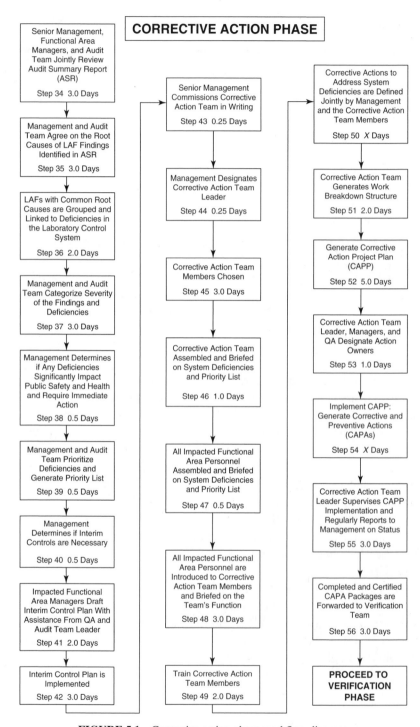

FIGURE 5.1   Corrective action phase workflow diagram.

TABLE 5.1   Explanation of Corrective Action Phase Workflow Diagram Steps

Step	Description	Estimated Duration	Explanation
34	Senior management, functional area managers, and audit team jointly review audit summary report (ASR)	3.0 days	Audit summary report (ASR), which was generated and circulated at the completion of the Reporting phase (see Chapter 4), is reviewed and discussed in a joint session. This session should include senior management, the functional area managers, and the members of the audit team. Discussion should be open and frank. The overall goal of these discussions is reach consensus as to the level of compliance which actually exists in all audited functional areas. This is accomplished by reviewing the summarized findings in the ASR. The allocation of 3 days may be insufficient to insure completed and open debate.
35	Management and audit team agree on the root causes of LAF findings identified in ASR	3.0 days	Chapter 4, shows how the ASR is organized by laboratory subelements. Within each section of the report is a table showing each gap (which is documented on a LAF), its impact on compliance with CGMPs, a potential root cause, and some suggestions for potential corrective actions recommended by the author of the LAF. It is important that management and the audit team agree on the root cause of the gap. This may take some debate in that root causes may be difficult to identify and may have far reaching significance.
			For example, say a finding shows repeated failures on the part of an analyst to properly obtain consistent results using a validated method. This may (and often is) attributed to lack of the analyst's training (a common scapegoat for bad methods). However, a careful root cause analysis may show that several analysts are having the same difficulty with the method. This in turn may suggest that the method does not perform as per its intended use and therefore, not properly validated. Further review may show that the method's validation SOP is not up to current industry standard and the method was not properly validated. Therefore, the

(Continued)

TABLE 5.1 *(Continued)*

Step	Description	Estimated Duration	Explanation
			root cause of the analyst's inability to properly use the method is not a problem with training, but a gap in the methods validation subelement. This type of in depth probing must be conducted by management and the audit team to insure the real cause (i.e., the root cause) of the problem can be addressed.
36	LAFs with common root causes are grouped and linked to deficiencies in the laboratory control system	2.0 days	Once all of the root causes for the gaps have been agreed upon, management and the audit team needs to group LAFs with common root causes. These findings should, in turn, have a common laboratory control system subelement system deficiency which resulted in their occurrence. If at all possible, gaps need to be linked to deficiencies in one of the seven laboratory control system subelements. If a systems-based approach to gaps designation and correction is not taken, there is significant risk of implementing Band-Aid solutions to the problems. This is the opposite of systems-based solutions which insure sustainable compliance.
37	Management and audit team categorize severity of the findings and deficiencies	3.0 days	As listed on the LAF, findings need to be categorized and assigned a priority: Immediate, high, routine, and low priority assignments indicate the severity of the gap. "Immediate" implies a significant violation of the CGMPs that could substantially impact the safety and efficacy of the finished product. "High" implies those gaps which could result in an FDA 483 finding or worse. "Routine" and "low" are gaps associated with administrative systems and practices, which by themselves, are not violations of the CGMPs but if not addressed can lead to failures. Although categorized by the author of the LAF, management and the audit team need to agree on the severity of the individual findings and deficiencies. (*Note*: The difference between a finding and deficiency is that a deficiency is a clear violation of 21 CFR parts 210 and 211 of CGMP regulations. A finding is a gap

TABLE 5.1 *(Continued)*

Step	Description	Estimated Duration	Explanation
			associated with administrative systems and practices, which by themselves, are not violations of CGMPs but if not addressed can lead to future deficiencies.)
38	Management determines if any deficiencies significantly impact public safety and health and requires immediate action	0.5 days	Deficiencies that rate an immediate classification must be clearly recognized by management. These actions will require immediate action on the part of the organization. Depending upon their severity, they may fall outside the normal corrective action system.
39	Management and audit team prioritize deficiencies and generate priority list	0.5 days	Gaps that are not immediate and do not pose a significant impact on public safety and health need to be prioritized. Management should create a priority list. This list indicates management's priorities for development and implementation of corrective and preventive actions.
40	Management determines if interim controls are necessary	0.5 days	Besides the immediate classification, certain gaps rate a high designation (e.g., which would result in action by FDA) and may require some interim controls to help mitigate their impact prior to formal corrective action. This also proves due diligence to FDA in the event it arrive on site prior to implementation of corrective and preventive actions. Senior management decide if actions require interim controls.
41	Impacted functional area managers draft interim control plan with assistance from QA and audit team leader	2.0 days	Functional area managers take those gaps, which were identified by senior management, to require interim controls and develop an interim control plan.
42	Interim control plan is implemented	3.0 days	Once developed and approved by senior management, the interim control plan is implemented. This plan also needs to be integrated into future corrective action plans.
43	Senior management commissions corrective action team in writing	0.25 days	This commissioning document should include the following sections: (1) Purpose, (2) Start date, (3) End date, (4) Expected

(Continued)

TABLE 5.1 (*Continued*)

Step	Description	Estimated Duration	Explanation
			deliverables, and (5) Definition of the team responsibilities, level of authority, and accountabilities. The commissioning document is signed and formally issued to the team leader once that individual is selected. Moreover, copies of the document should be circulated to all impacted personnel within the organization. The commissioning of the corrective action team should be a well-publicized event, as was the audit.
44	Management designates corrective action team leader	0.25 day	The corrective action team leader is held accountable for the successful development of the corrective action project plan and subsequent development and implementation of corrective and preventive actions. An individual with good project management and organizational skills is required. Although quality assurance personnel are often considered for such roles, laboratory managers and supervisors should be considered as well.
45	Corrective action team members chosen	3.0 days	The corrective action team should include as many personnel from within the impact functional areas as possible. A much greater likelihood of success exists if bench-level personnel are actively involved in upgrading the level of compliance within their sections, groups, or departments. The actual size and composition of the corrective action team is determined by the audit findings. Team composition need not be rigid but can be modified as needed.
46	Corrective action team assembled and briefed on system deficiencies and priority list	1.0 day	Team members review the gap priority list and corresponding system deficiencies that generated the findings.
47	All impacted functional area personnel assembled and briefed on system deficiencies and priority list	0.5 day	Functional area personnel review the gap priority list and corresponding system deficiencies that generated the findings. All area personnel should be aware of the deficiencies and their root causes.

TABLE 5.1 (*Continued*)

Step	Description	Estimated Duration	Explanation
48	All impacted functional area personnel are introduced to corrective action team members and briefed on the team's function	3.0 days	A review of the commissioning document is appropriate at this stage. Again, this should include the following sections: (1) Purpose, (2) Start date, (3) End date, (4) Expected deliverables, and (5) Definition of the team responsibilities, level of authority, and accountabilities.
49	Train corrective action team members	2.0 days	At a minimum, training should include: (1) Review of the goals of the audit, (2) Review of the entire audit process, (3) Review of roles and responsibilities, (4) Discussion of the tentative working calendar, and (5) An overview of the LAF-to-corrective and preventive action workflow. A discussion with respect to systems-based compliance with CGMPs is included.
			A generic diagrammatical description of the LAF-to-corrective-and-preventive action workflow process is shown in Figure 5.2.
50	Corrective actions to address system deficiencies are defined jointly by management and the corrective action team members	X days	Following identification of LAFs with common root causes and correlation of these findings to deficiencies in the Laboratory Control System, corrective actions need to be developed to eliminate the deficiencies in the systems. Since the potential impact of these corrective actions is long term and far reaching, management at all levels and the corrective action team members jointly define the actions. The defined actions are broad yet specifically address the system deficiencies. Details of how these actions are implemented is left up to the corrective action team.
			The time needed to complete this step is dependant upon the number of LAFs and corresponding system deficiencies. Therefore, no duration is assigned.
51	Corrective action team generates work breakdown structure	2.0 days	As a first step to address the corrective actions defined in Step 50, the corrective action team generates a work breakdown structure. A work breakdown structure

(*Continued*)

TABLE 5.1   (*Continued*)

Step	Description	Estimated Duration	Explanation
			(WBS) is created by a group brainstorming session that identifies the steps required to implement the corrective actions.
52	Generate corrective action project plan (CAPP)	5.0 days	The corrective action project plan is created from the WBS. The project plan builds on the steps identified in the work breakdown structure. However, additional information such as start and stop dates and milestones are included. The use of project management software is acceptable but not necessary to generate the project plan.
53	Corrective action team leader, managers, and QA designate action owners	1.0 day	Once the project plan is complete, the corrective action team works with the managers and quality assurance to assign ownership to the actions. The corrective action team assists in the execution of the corrective action project plan and subsequent implementation of corrective and preventive actions, but individuals outside the team are ultimately responsible for remediation of individual actions. The selection of bench personnel and supervisors as action owners is encouraged.
54	Implement CAPP: Generate corrective and preventive actions (CAPAs)	X days	Once the action owners are assigned, the corrective action project plan is executed. Action owners with the assistance of the corrective action team members execute the project plan and complete the corrective and preventive actions defined in the project plan.
55	Corrective action team leader supervises CAPP implementation and regularly reports to management on status	3.0 days	As with any project, planning and supervision are critical to successful completion of the plan. The corrective action team should report at least monthly to management on the status of the completion of the corrective actions.
56	Completed and certified CAPA packages are forwarded to verification team	3.0 days	When corrective and preventive actions have been completed, supporting documentation, including in-use data, are forwarded to verification team. The functions of the verification team are described in Chapter 6.

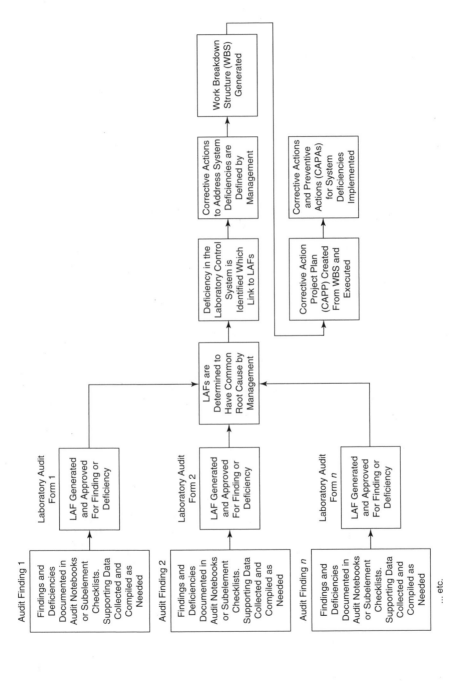

FIGURE 5.2 Generic LAF-to-CAPA workflow diagram.

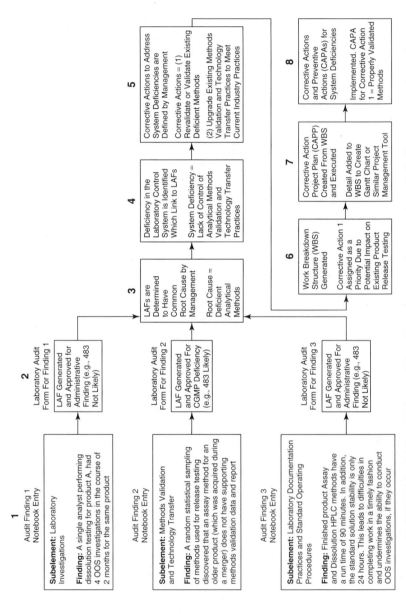

1

Audit Finding 1
Notebook Entry

Subelement: Laboratory Investigations

Finding: A single analyst performing dissolution testing for product A, had 4 OOS investigations in the course of 2 months for the same product

2

Laboratory Audit Form For Finding 1

LAF Generated and Approved for Administrative Finding (e.g., 483 Not Likely)

Audit Finding 2
Notebook Entry

Subelement: Methods Validation and Technology Transfer

Finding: A random statistical sampling of methods used for release testing discovered that an assay method for an older product (which was acquired during a merger) does not have supporting methods validation data and report

Laboratory Audit Form For Finding 2

LAF Generated and Approved For CGMP Deficiency (e.g., 483 Likely)

Audit Finding 3
Notebook Entry

Subelement: Laboratory Documentation Practices and Standard Operating Procedures

Finding: Finished product Assay and Dissolution HPLC methods have a run time of 90 minutes. In addition, the standard solution stability is only 24 hours. This leads to difficulties in completing work in a timely fashion and undermines the ability to conduct OOS investigations, if they occur

Laboratory Audit Form For Finding 3

LAF Generated and Approved For Administrative Finding (e.g., 483 Not Likely)

3

LAFs are Determined to Have Common Root Cause by Management

Root Cause = Deficient Analytical Methods

4

Deficiency in the Laboratory Control System is Identified Which Link to LAFs

System Deficiency = Lack of Control of Analytical Methods Validation and Technology Transfer Practices

5

Corrective Actions to Address System Deficiencies are Defined by Management

Corrective Actions = (1) Revalidate or Validate Existing Deficient Methods

(2) Upgrade Existing Methods Validation and Technology Transfer Practices to Meet Current Industry Practices

6

Work Breakdown Structure (WBS) Generated

Corrective Action 1 Assigned as a Priority Due to Potential Impact on Existing Product Release Testing

7

Corrective Action Project Plan (CAPP) Created From WBS and Executed

Detail Added to WBS to Create Gantt Chart or Similar Project Management Tool

8

Corrective Actions and Preventive Actions (CAPAs) for System Deficiencies

Implemented. CAPA for Corrective Action 1 = Properly Validated Methods

FIGURE 5.3 LAF-to-CAPA workflow diagram: converting example audit findings to corrective and preventive actions. For discussion, see p. 66.

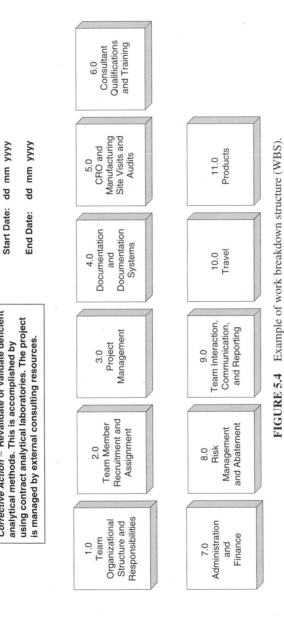

FIGURE 5.4 Example of work breakdown structure (WBS).

Corrective Action = Revalidate or validate deficient analytical methods. This is accomplished by using contract analytical laboratories. The project is managed by external consulting resources.

Start Date: dd mm yyyy

End Date: dd mm yyyy

1.0 Team Organizational Structure and Responsibilities

2.0 Team Member Recruitment and Assignment

3.0 Project Management

4.0 Documentation and Documentation Systems

5.0 CRO and Manufacturing Site Visits and Audits

6.0 Consultant Qualifications and Training

7.0 Administration and Finance

8.0 Risk Management and Abatement

9.0 Team Interaction, Communication, and Reporting

10.0 Travel

11.0 Products

1.0 Team Organizational Structure and Responsibilities

- 1.1 Define Team Objectives
- 1.2 Identify Existing/Needed Team Member Skill Sets
- 1.3 Define Team Structure
- 1.4 Define Team Members Responsibilities
- 1.5 Define Team Member Skill Overlap
- 1.6 Address Team Member Time Overlaps

2.0 Team Member Recruitment and Assignment

- 2.1 Create Team Member Roles and Responsibilities Matrix
- 2.2 Fill in Draft Org. chart
- 2.3 Review and Critique Draft Org. Chart
- 2.4 Review and Critique Draft Org. Chart with Client: Get approval
- 2.5 Prepare Team Organization and Integration Presentation
- 2.6 Team Meeting and Organization and Integration Session

3.0 Project Management

- 3.1 Assign Project Manager
- 3.2 Assign Assistant/Backup Project Manager
- 3.3 Define Deliverables : Validation Review Packages and Tech Transfer Packages
- 3.4 Review Current Client Processes
- 3.5 Draft Process Diagram for Completing Deliverables
- 3.6 Estimate Time Cycle for Steps in Process Diagram
- 3.7 Brainstorm Draft with Team Members
- 3.8 Assign Project-to-Project Managers
- 3.9 Address Product Adds and Drops
- 3.10 Choose Software Tools With Client Input

4.0 Documentation and Documentation Systems

- 4.1 Define What SOPs Will Impact Project From Quality Systems Perspective
- 4.2 Obtain Client Master SOP List: ID Existing Client SOPs That Impact Project
- 4.3 Obtain Contract Labs SOP List: ID SOPs That Impact Project
- 4.4 Perform GAP Analysis on Additional Needed SOPs
- 4.5 Define System for Consultant Modification of SOPs
- 4.6 Modify, Obtain, Write, or Have Written SOPs ID'd in Gap Analysis
- 4.7 ID Parties Who Will Need to Share Documents

5.0 CRO and Manufacturing Site Visits and Audits

- 5.1 Determine What Audits/ Visits are Needed: Initial, Periodic, and Final
- 5.2 Determine Difference Between Visit and Audit
- 5.3 Determine Sites In Which Audits/Visits Need to Be Conducted
- 5.4 Develop Audit/Visit Checklists
- 5.5 Get Visit/Audit Approval and Participation Desires From Client
- 5.6 Obtain Site Contacts to Arrange Visits/Audits
- 5.7 Establish Timeline for Visits/Audits
- 5.8 Integrate Visits/Audits Into Project Plan
- 5.9 Define Audit/Visit Completion Report
- 5.10 Execute Visits/Audits
- 5.11 Publish Audit/Visit Report: Determine if Corrective Action is Needed

FIGURE 5.4 (Continued)

Laboratory Control System Deficiency: There is a lack of control of analytical methods validation and technology transfer practices at the site.

Corrective Action 1: Revalidate or validate deficient analytical methods. This will be accomplished by using contract analytical laboratories. The project will be managed by external consulting resources.

Step 3.0 Project Management

ID	Task Name	Start	End	Duration
1	3.1 Assign Project Manager	4/14/2008	4/15/2008	2d
2	3.2 Assign Assistant/Backup Project Manager	4/16/2008	4/17/2008	2d
3	3.3 Define Deliverables: Validation Review Packages and Tech Transfer Packages	4/18/2008	4/23/2008	4d
4	3.4 Review Current Client Processes	4/15/2008	4/15/2008	1d
5	3.5 Draft Process Diagram For Completing Deliverables	4/23/2008	4/25/2008	3d
6	3.6 Estimate Time Cycle for Steps in Process Diagram	4/23/2008	4/25/2008	3d
7	3.7 Brainstorm Draft with Team Members	4/28/2008	4/29/2008	2d
8	3.8 Assign Project to Project Managers	4/18/2008	4/18/2008	1d
9	3.9 Address Product Adds and Drops	4/14/2008	4/14/2008	1d
10	3.10 Choose Software Tools With Client Input	5/1/2008	5/2/2008	2d
11	3.11 etc.	4/30/2008	5/6/2008	5d
12				

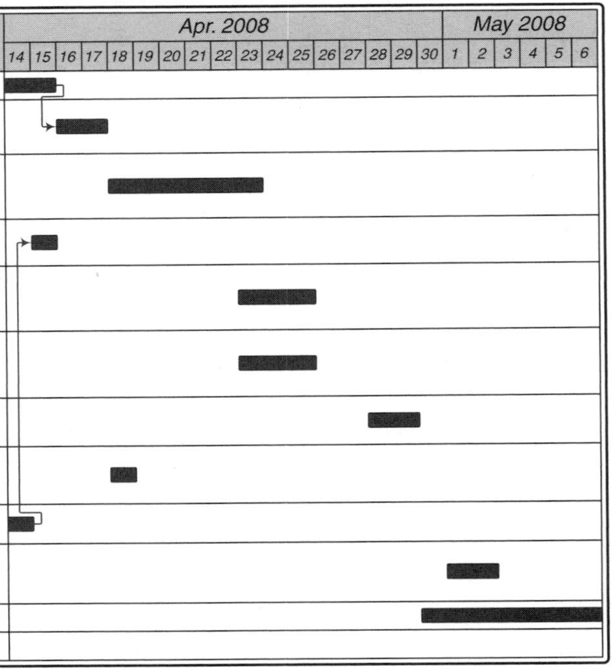

FIGURE 5.5 Example of project plan in Gantt chart format.

5.2 LAF-TO-CAPA WORKFLOW DIAGRAM: CONVERTING EXAMPLE AUDIT FINDINGS TO EXAMPLE CORRECTIVE AND PREVENTIVE ACTIONS

The following is an analysis of steps shown in Figure 5.3.

5.2.1 Step 1 Audit Finding Notebook Entries

For this example, three audit notebook entries where made with respect to findings during audit of the laboratory investigations, methods validation, and laboratory documentation and standard operating procedures subelements. The observations made and recorded in the notebooks were:

Audit Finding 1 Notebook Entry

Subelement: Laboratory investigations

Finding: A single analyst performing dissolution testing for product A, had 4 OOS investigations in the course of 2 months for the same product.

Audit Finding 2 Notebook Entry

Subelement: Methods validation

Finding: A random statistical sampling of methods used for release testing discovered that an assay method for an older product (which was acquired during a merger) does not have supporting methods validation data and report.

Audit Finding 3 Notebook Entry

Subelement: Laboratory documentation and standard operating procedures

Finding: Finished product assay and dissolution HPLC methods have a run time of 90 minutes. In addition, the standard solution stability is only 24 hours. This leads to difficulties in completing work in a timely fashion and undermines the ability to conduct OOS investigations if they occur.

Although these findings do not initially appear to be linked, further analysis shows otherwise.

5.2.2 Step 2 Formal Documentation of Finding or Deficiency on LAFs

As previously described, notebook findings are formally documented on LAFs. In this example, notebook entry 1, would not likely rate a Form 483 observation by FDA and therefore is listed as an administrative finding. Notebook entry 2 is a clear violation of CGMPs and would rate a Form 483 citation and is therefore, a deficiency. Notebook entry 3 is an administrative finding.

5.2.3   Step 3 Common Root-Cause Correlation by Management

During review of the audit summary report (see Step 36 in Table 5.1), management will group LAFs with common root causes. This root cause assignment may not necessarily correspond to the LAF author's initial assignment, due to the fact that once completed, the ASR gives a broader perspective of the potential system deficiencies. In this case, three distinctly different LAFs have been assigned the same root cause by management, namely, the fact that certain analytical methods are deficient.

For example, review of LAF 1 data showed that the analyst was very well-trained and the method appeared to be the root cause. For LAF 2, the method was deemed to be suspect because no data supported that it functions properly for intended use. For LAF 3, the extended run times and lack of standard solution stability are clearly problematic. These examples show the subtleties associated with root cause analysis and the level of care that needs to be taken in assignment of root causes.

5.2.4   Step 4 LAF Linkage to System Deficiencies

Further analysis (for this example) showed that the methods associated with all three LAFs were deficient as a result of improper or nonexistent methods validation procedures. Therefore, management determined that the system deficiency for these three findings was a lack of control of analytical methods validation and technology transfer practices at the site.

5.2.5   Step 5 Management Assignment of Corrective Actions to Address System Deficiency

To address this system deficiency management determined the need for two corrective actions:

1. Revalidate or validate existing deficient methods
2. Upgrade existing methods validation and technology transfer practices to meet current industry standards

Note that the actions are worded in very general or high level language. This is done on purpose to highlight the system deficiency and avoid addressing specific actions thus resulting in a Band-Aid approach.

To add further detail for this example, the action will read: "Revalidate or validate deficient analytical methods. This will be accomplished by using contract analytical laboratories. The project will be managed by external consulting resources."

5.2.6   Step 6 Work Breakdown Structure (WBS) Is Generated

A work breakdown structure (WBS) is created by a corrective action team group brainstorming session to identify the steps required to implement the corrective actions. It starts with identifying those major work functions that are part of the corrective action. Detailed steps are added once consensus is reached on the major work functions. The WBS is a rough out of the final project plan, and does not contain details such as start and stop dates, resources required, and so on.

Figure 5.4 shows a partial WBS example for corrective action: "Revalidate or validate deficient analytical methods. This will be accomplished by using contract analytical laboratories. The project will be managed by external consulting resources."

5.2.7   Step 7 Corrective Action Project Plan (CAPP) Created From WBS and Executed

The corrective action project plan is created from the WBS. The project plan builds on the steps identified in the WBS. However, additional information such as resource assignment, start and stop dates, and milestones are included. The use of project management software is acceptable but not necessary to generate the project plan. Figure 5.5 shows a partial project plan in Gantt chart format.

5.2.8   Step 8 Corrective and Preventive Actions (CAPAs) for System Deficiencies

Following completion of the project plan, corrective and preventive actions should be in place. Moreover, supporting data showing the steps taken to implement the actions as well as in-use data showing the successful use of the upgraded systems are included in the final data package. These data are combined with the original LAFs and audit findings to form a complete life-cycle history of the discovery of the finding to its successful completion. These data are then forwarded to the verification team.

CHAPTER 6

DEVELOPING AND IMPLEMENTING A VERIFICATION PLAN

6.1 PROCEDURE

In order to ensure that the corrective action project plan (CAPP) is being properly implemented and that the corrective and preventive actions (CAPAs) are leading to sustainable compliance, a verification plan needs to be developed and implemented. The verification plan execution is monitored by a verification review board and corrective actions are verified by a verification team. The steps in this process are shown Figure 6.1 and described in Table 6.1.

6.2 CORRECTIVE ACTION VERIFICATION PROCESS

Figure 6.2 is a graphical representation of the corrective action verification process. A detailed explanation of each of the steps shown in Figure 6.2 (p. 76) is given in the following text.

Establishing a CGMP Laboratory Audit System. By David M. Bliesner
Copyright © 2006 John Wiley & Sons, Inc.

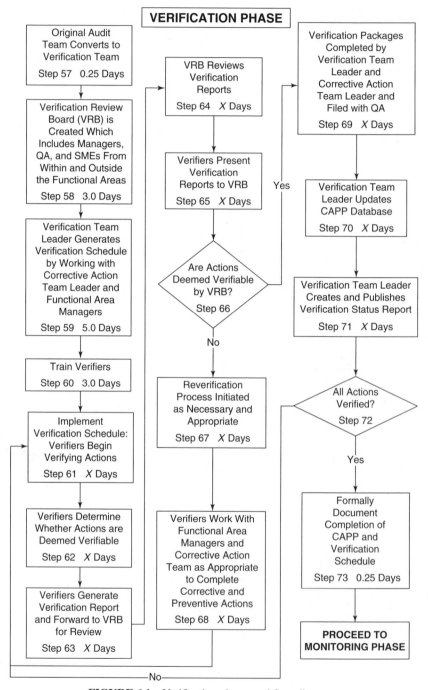

FIGURE 6.1   Verification phase workflow diagram.

TABLE 6.1 Explanation of Verification Phase Workflow Diagram Steps

Step	Description	Estimated Duration	Explanation
57	Original audit team converts to verification team	0.25 day	The verification team is responsible for verifying or confirming that corrective and preventive actions have been properly designed and implemented. The most effective and efficient means of verifying implementation of corrective and preventive actions is to let the personnel who originally found the gaps verify that they have been corrected. By doing this, less of a chance of duplication of efforts exists, in that the original audit team members are already familiar with the gaps and corresponding system deficiency. Therefore, if possible, the audit team should assume the responsibility of the verification team once the Verification phase has begun. Verification team members are normally referred to as verifiers. The record of their verification activities are documented in a verification report.
58	Verification review board (VRB) is created which includes managers, QA, and SMEs from within and outside the functional areas	3 days	The VRB reviews the verification reports generated by the verification team members and determines whether it agrees with the recommendations of the report. A verification review board is similar to a quality council in that its main mission is to insure that systems based corrective and preventive actions have been implemented by the action owners with the help of the corrective action team. The team is composed of a diverse group of personnel from within the organization and needs not be formed from personnel who are all laboratory oriented. In many cases, it is quite valuable to have nonlaboratory personnel function as VRB members because they bring a different perspective on CGMP compliance. In addition, board participation is an excellent means to familiarize nonlaboratory personnel with laboratory compliance responsibilities. Managers, quality assurance personnel and subject matter experts are some recommendations for board member composition.

(Continued)

TABLE 6.1 (*Continued*)

Step	Description	Estimated Duration	Explanation
59	Verification team leader generates schedule by working with corrective action team leader and functional area managers	5 days	The verification team leader, who was previously the audit team leader, needs to work closely with the action owners, the corrective action team leader, and the functional area managers to determine when actions are ready to be verified. Actions are deemed to be verifiable or ready to be verified, when the specific gap has been corrected and the corresponding system deficiency has been repaired. Moreover, the corrective and preventive action should be in place long enough so supporting data shows that the implemented actions are effective. These data are referred to as in-use data. In some circumstances, a significant amount of time is required before an action has enough in-use data to be deemed verifiable. In addition, some corrective and preventive actions are linked and require completion of several or all of the linked actions before any one action can be verified.
			In creating the verification schedule, the verification team leader insures that these data exist and that any linkages are clearly identified with their supporting data so that the verifier's time is not wasted. The verification schedule can be based on the corrective action project plan except later in time.
60	Train verifiers	3 days	As with auditing, it is important to train verifiers on the verification process. Figure 6.2 (see p. 76) shows what the process verifiers need to be familiar with in order to insure that proper corrective and preventive actions have been implemented.
61	Implement verification schedule: Verifiers begin verifying actions	X days	Verifiers should prepare a verification plan prior to verifying any actions. An example verification plan is shown in Figure 6.3 (see p. 79). Plans may be reviewed by the verification team leader who in turn may wish to share the plan with the action owners and functional area managers.
			Number of days required to complete this task and Steps 62–69 depend on the findings

TABLE 6.1   (*Continued*)

Step	Description	Estimated Duration	Explanation
			and actions developed for a particular site. Therefore, no time estimate is given.
62	Verifiers determine whether actions are deemed verifiable	X days	There are usually three outcomes that result from the verification of the process: (1) Action is verified as complete, (2) Action is not verified as complete, with the deficiencies clearly defined, and (3) Action is verifiable but is pending sufficient in-use data. The verifiers determine this status based on the supporting data presented by the action owners. This is the recommendation made to the verification review board and defended in front of the board.
63	Verifiers generate verification report and forward to VRB for review	X days	Following the determination of the status of action verification, the verifiers generate a corrective action verification report. This report should be simple, yet be able to stand on its own in support of the verifier's position with respect to the status of the action verification.
64	VRB reviews verification reports	X days	Review board members read the verification reports prior to presentation by the verifiers. Since this report may be used in the future to support the company's position to an outside organization with respect to compliance, the board members look at the document to see whether it can stand on its own without additional input from the verifier. A critical review is most necessary in preparation for the interview with the verifier.
65	Verifiers present verification reports to VRB	X days	During this step, verifiers review the corrective action verification report with the review board and defend their position. The presentation before the review board is a rigorous process in that this is the organization's collective decision that the gap has been closed and that the deficiency in the system which has created that gap has been corrected. The review board may accept or reject the verifiers' recommendation. Although this may seem like overkill, it insures that the entire organization is onboard with the decision

(*Continued*)

TABLE 6.1 (*Continued*)

Step	Description	Estimated Duration	Explanation
			that the corrective action is sufficient and the corrective and preventive actions result in long-term or sustainable compliance with CGMPs.
66	Actions deemed verifiable by VRB	*X* days	The board concurs or disagrees with verifiers' conclusion that: (1) Action is verified as complete, (2) Action is not verified as complete, with the deficiencies clearly defined, and (3) Action is verifiable but is pending sufficient in-use data.
67	Reverification process initiated as necessary and appropriate	*X* days	Those actions which are deemed not verifiable or verifiable pending additional in-use data are recycled back through a reverification process. The amount of rework depends upon the board's recommendations as to what needs to be completed prior to agreeing that the action is verifiable. The board should be very clear in defining what additional evidence is needed to complete corrective actions which require additional in-use data.
68	Verifiers work with functional area managers and corrective action team as appropriate to complete corrective and preventive actions	*X* days	Following the board's ruling and direction, the verifiers, functional area managers, and corrective action team members work closely with the action owners to insure the appropriate steps are taken to complete the corrective action.
69	Verification packages completed by verification team leader and corrective action team leader and filed with QA	*X* days	Once the actions have been deemed verifiable by the verification review board, it is time for them to be officially closed. The action owner, corrective action team leader, and verification team leader insure that verifiable corrective actions are supported by the necessary data. The finished product of a verifiable action should be a data package with the original LAF as the cover sheet, the supporting audit data, the categorization and root cause of the finding or deficiency defined by management, the corrective action to address the system deficiency defined jointly by management and the corrective action team members, the verification plan, the corrective action verification report, and all supporting

TABLE 6.1   (*Continued*)

Step	Description	Estimated Duration	Explanation
			verification data. This package is proof to any reviewer of the concerted effort and following successful implementation of a laboratory control system audit process.
70	Verification team leader updates CAPP database	0.25 day	The verification team leader makes appropriate entries into the database indicating that the corrective and preventive actions have been successfully implemented.
71	Verification team leader creates and publishes verification status report		After the database entries are made, the verification team leader creates and publishes a verification status report and circulates it to all appropriate personnel. This document should be published frequently to insure timelines are maintained and deadlines are met.
72	All actions verified?		If all actions are not verified, the verification schedule is maintained and the verification process continues.
73	Formal document completion of CAPP and verification schedule		If all actions are verified and all supporting data packages have been completed and filed, a formal memorandum is generated by the verification team and submitted to senior management and QA for review and approval. This information is shared with the entire organization indicating the successful completion of the audit and verification process. The organization now enters a monitoring phase to insure sustainable compliance and to prevent recidivism.

6.2.1 Step 1 Action Owners Work with Corrective Action Team to Design and Implement Systems-Based Corrective Actions

Recall that corrective actions to address system deficiencies are defined jointly by management and the corrective action team members. These actions are subsequently assigned to an action owner or owners who are responsible for developing and implementing corrective and preventive actions with the assistance of the corrective action team. These actions may address specific gaps but should also correct the system deficiencies that lead to their creation. See Figure 6.2.

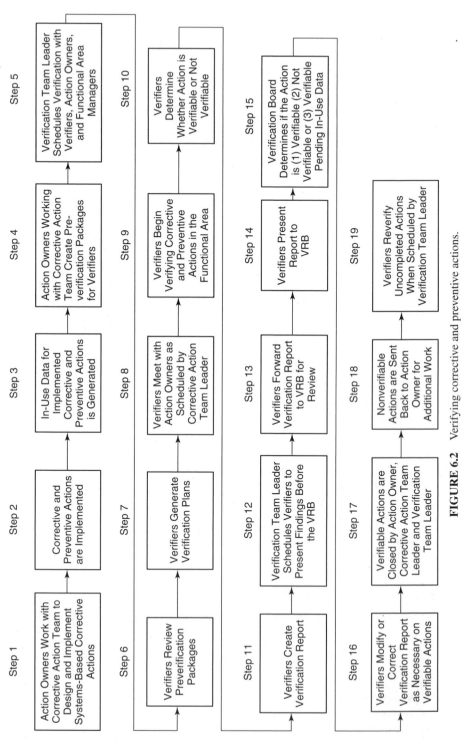

FIGURE 6.2 Verifying corrective and preventive actions.

6.2.2   Step 2 Corrective and Preventive Actions Are Implemented

Execution of the corrective action project plan should lead to implementation of sustainable corrective and preventive actions.

6.2.3   Step 3 In-Use Data for Implemented Corrective and Preventive Actions Are Generated

Following implementation of the corrective and preventive actions, in-use data will be generated showing the systems based corrective and preventive actions are actually working. This may take some time. For example, if the corrective action involves implementing a formalized laboratory training and qualification system, it may take in excess of 6 months to hire a laboratory trainer, generate a laboratory training and qualification SOP, and development of training and qualification materials to support the SOP. In turn, it may take an additional 6 months or more to generate enough in-use data to show that the new training system is in place and working as designed. In-use data in this example would include proof of hiring, an approved training and qualification SOP, and completed training records showing the retrospective training of personnel as defined in the new SOP. In addition, training of newly hired personnel under the new SOP should be part of the supporting in-use data.

6.2.4   Step 4 Action Owners Working with Corrective Action Team Create Preverification Packages for Verifiers

Once action owners and their managers are confident they have completed the corrective and preventive actions and have sufficient in-use data to support their positions, they work with the corrective action team to generate supporting data packages. These data packages are sometimes referred to as preverification packages and are forwarded to the verifiers prior to the actual verification action. These packages should be as comprehensive as possible so enough information is provided for the verifiers to develop verification plans limiting the amount of time needed during actual verification efforts.

6.2.5   Step 5 Verification Team Leader Schedules Verification with Verifiers, Action Owners, and Functional Area Managers

Much of the verification team leader's time is occupied with schedule action verification. All the parties involved with verification including the action owners, verifiers, and functional area managers and supervisors need to be in agreement that the action is ready to be verified. If not, everyone's time runs the risk of being wasted. If the action is not ready to be verified, the owner must convey this to all involved parties and reschedule with the verification team leader.

6.2.6   Step 6 Verifiers Review Preverification Packages

The verifiers are provided with preverification packages and should review them prior to generation of a verification plans and action verification. These packages include plans, protocols, and supporting data demonstrating that the corrective action was taken.

6.2.7   Step 7 Verifiers Generate Verification Plans

In order to insure smooth and efficient action verification, it is recommended that the verifier create a verification plan. Figure 6.3 shows an example corrective action verification plan. This plan is reviewed by the verification team leader and shared with the action owner prior to verification of the action.

6.2.8   Step 8 Verifiers Meet with Action Owners as Scheduled by Corrective Action Team Leader

Prior to the actual verification, the verifier should meet face-to-face with the action owner and review the verification plan and any modifications to the plan they intend to take. A preverification meeting, just prior to the action verification, insures efficient use of time and resources.

6.2.9   Step 9 Verifiers Begin Verifying Corrective and Preventive Actions in the Functional Area

The verifiers begin the action verification by executing the verification plan. Verification should be thorough and complete, in order to insure that there are data to support the position of verifiable, non-verifiable, and verifiable pending in-use data. Verifiers must keep in mind that they will have to support their recommendations in front of the verification review board. This is somewhat akin to defending a thesis, both in rigor and style. As with auditing, verification should involve at least two people; one asking questions and the second taking notes.

6.2.10   Step 10 Verifiers Determine Whether Action Is Verifiable or Not Verifiable

As stated in Step 9, there are three possible outcomes: verifiable, nonverifiable, and verifiable pending in-use data. An additional position is possible, action not ready for verification, but this should rarely be the case if all parties are actively engaged in the corrective action process.

Corrective Action Verification Plan

Site: Yours   **Action No.** xxxxxx   **Department:** Theirs   **Due Date:** 9/28/08

Corrective Action: Review All HPLC Methods to Confirm They Have Been Validated for Their Intended Use.

Background: During the original audit, several observations were made with respect to the suitability of the methods for their intended use. Some of these findings included:

- Some methods appear not to have not been validated under current guidelines but are being used to perform release for finished products.
- Some methods are being used in analyses where stability indication is important and they are not stability-indicating methods.
- Portions of method robustness, including determining the effect of HPLC column temperature variation on method performance, are not performed during methods validation.

Scope: Verification will encompass all HPLC analytical methods used in Their Department for all approved products. Specifically, verification will entail a review of statistical sampled HPLC analytical methods to determine if they are: (1) Stability indicating as appropriate and (2) Temperature robustness was performed during methods validation. A subsequent review of the verification documentation shows if these methods were reviewed as part of the corrective action and whether they have been properly validated for their intended use. A review of plans to revalidate or validate deficient methods is also be performed.

Documents to Review:
1. Master HPLC analytical methods list for all approved products within Their Department.
2. Current revision of SOP ABC-001-001 "Validation of Analytical Procedures".
3. Written plans and/or protocols delineating a systematic approach for review of these methods during the corrective action process.
4. Original methods validation reports for sampled methods, and the subsequent protocols and reports of revalidated methods.

Planned Interviews/Observations of Action In Use:
1. Interview selected personnel who were involved in the methods review and/or revalidation effort to evaluate understanding of the roles and responsibilities in the process.
2. Interview management to determine how the plans and protocols were implemented.
3. Review report associated with review of all analytical methods and subsequent conclusions on the appropriateness of their supporting validations.

FIGURE 6.3   An example of a corrective action verification plan.

6.2.11   Step 11 Verifiers Create Verification Report

Following the determination of the status of action verification, the verifiers generate a corrective action verification report. This report should be simple, yet be able to stand on its own in support of the verifiers' position with respect to the status of the action verification. An example corrective action verification report is shown in Figure 6.4.

Corrective Action Verification Report

Site: Yours   **Action No.** xxxxxx   **Department:** Theirs   **Due Date:** 09/28/08

Corrective Action: Review All HPLC Methods to Confirm They Have Been Properly Validated for Their Intended Use

Background and Approach:

During the original audit, several observations were made with respect to the suitability of the methods for their intended use. Some of these findings included:

- Some methods appear not to have not been validated under current guidelines but are being used to perform release for finished products.
- Some methods are being used in analyses where stability indication is important and they are not stability indicating methods.
- Portions of method robustness, including determining the effect of HPLC column temperature variation on method performance, are not performed during methods validation.

The verification was performed 9/25/2008. The following steps were taken during verification:

- Verification entailed a review of statistical sampled HPLC analytical methods (e.g., 9 product files were randomly sampled and 3 products were reviewed) to determine if they are:

  - Stability indicating as appropriate
  - Temperature robustness was performed during methods validation

- A subsequent review of the verification documentation showed these methods were reviewed as part of the corrective action and have been properly validated for their intended use.
- Plans exist to revalidate or validate deficient methods.
- The method review conclusions were appropriate and accurate.

Verifier(s)/Verification Date(s):

T.J. Booker, D. Blistex, 25 September 2008

Conclusion: Verified? Yes ☒ No ☐

Support for Conclusion

1. One document titled, "Corrective Action Plan for Method and Methods Validation Report Review" exists. This document provides definitions of temperature robustness and stability indication and delineates the steps required to complete the action.

2. On 9/25/2008, one document titled, "Review of The Validation Status of HPLC Methods" was reviewed by the verifiers. This document provides evidence that the action plan was completed.

3. Random statistical sampling and assessment of methods validation reports showed that the conclusions reached in the departmental review were appropriate and their reports accurate.

4. Plans exist to generate protocols to validate methods which were found to be deficient in their supporting methods validation data.

FIGURE 6.4 An example of a corrective action verification report.

Supporting Documentation:

Documents Reviewed (SOP, Report, Training Materials, etc.)	Document Date and Version
Corrective Action Plan for Method and Methods Validation Report Review	08/15/07 rev0
Review of The Validation Status of HPLC Methods	10/15/07 rev0
SOP ABC-001-001 Validation of Analytical Procedures	02/02/07 rev6
STM-001 Determination of Ibuprofen in Ibuprofen Tablets, USP 200 mg and 800 mg	04/16/05 rev3
RPT-001 Validation of the HPLC Assay Method for the Quantitative Determination of Ibuprofen and its Related Compounds in Ibuprofen Tablets USP 200 mg and 800 mg	07/27/04 rev0
RPT-002 Validation of the HPLC Assay Method for the Quantitative Determination of Hydrocodone and Its Related Compounds in Hydrocodone Tablets (25 mg and 50 mg)	08/05/06 rev0
STM-009 HPLC Method for the Assay of Hydrocodone in 25 mg and 50 mg Tablets	10/30/07 rev1
STM-005 HPLC Method for Assay, Content Uniformity, Blend Uniformity and Chromatographic Purity Testing of 5 mg Muctane HCl Tablets	07/18/04 rev5
RPT-005 Validation of the HPLC Assay, Content Uniformity, Blend Uniformity, and Chromatographic Purity Testing of 5 mg Muctane HCL Tablets	07/15/03 rev0

Comments if needed:

None.

Interviews:

Person/Title of Interviewee	Date
Dr. P. Sarbanes, Director Analytical Development	08/25/08
Ms. Tina Luis, Associate Analytical Development	08/25/08

Comments if needed:

None.

Meetings/Events attended:

Event Description, attendees/audience	Date
	Date of event
None	na

Comments if needed:

None.

Verifiers Name: T.J. Booker, D. Blistex   **Date:** _____

Verification Review Board Approval: _____ **Date:** _____

FIGURE 6.4 (*Continued*)

6.2.12   Step 12 Verification Team Leader Schedules Verifiers to Present Findings Before the Verification Review Board

Following verification, the verification team leader reviews the corrective action verification report and determines whether the verifiers have enough data to support their position in front of the verification review board. If verification team leader determines more data are needed, the verification continues. If not, the team leader schedules the verifier to present the report before the verification review board.

6.2.13   Step 13 Verifier Forwards Verification Report to Verification Review Board for Review

A copy of the corrective action verification report is forwarded to all the members of the verification review board for their review prior to the verifier's presentation and defense of position.

6.2.14   Step 14 Verifier Presents Report to Verification Review Board

During this step the verifiers review the corrective action verification report with the review board and defend their position. The presentation before the review board is a rigorous process in that this is the organization's collective decision that the gap has been closed and that the deficiency in the system which created that gap has been corrected. The review board may accept or reject the verifiers' recommendation. Although this may seem like overkill, it insures that the entire organization is onboard with the decision that the corrective action is sufficient and the corrective and preventive actions results in long-term or sustainable compliance with CGMPs.

6.2.15   Step 15 Verification Board Determines if the Action Is (1) Verifiable, (2) Not Verifiable or, (3) Verifiable Pending In-Use Data

Actions which are not verifiable or verifiable pending in-use data go into a do loop (e.g., work through the corrective action–verification cycle as many times as necessary to correct the problem). The reverification time frame is supervised by the verification team leader.

6.2.16   Step 16 Verifier Modifies or Corrects Verification Report as Necessary on Verifiable Actions

For those actions which are deemed verifiable, the verification review board may make suggestions for additions or improvements to report. Therefore, verifiers may have to make modifications to the report as necessary to serve the board's wishes.

6.2.17   Step 17 Verifiable Actions are Closed by Action Owner, Corrective Action Team Leader, and Verification Team Leader

The action owner, corrective action team leader, and verification team leader insure that verifiable corrective actions are supported by the necessary data. The finished product of a verifiable action is a data package which has the original LAF as the cover sheet, the supporting audit data, the categorization and root cause of the finding or deficiency defined by management, the corrective action to address the system deficiency defined jointly by management and the corrective action team members, the verification plan, the corrective action verification report, and all supporting verification data. This package is proof to any reviewer of the concerted effort and following successful implementation of a laboratory control system audit process.

6.2.18   Step 18 Nonverifiable Actions Are Sent Back to Action Owner for Additional Work

All nonverifiable or verifiable pending in-use data actions are forwarded back to the action owner and enter the do-loop cycle.

6.2.19   Step 19 Verifiers Reverify Uncompleted Actions When Scheduled by Verification Team Leader

If deemed not verifiable, the verification team leader begins working with the action owners, the functional area managers, and the corrective action team to correct any deficiencies or obtain additional supporting data to make the rejected actions verifiable. Once scheduled for reverification, the verifiers go through the same process as before to present their reports to the verification review board.

CHAPTER 7

DEVELOPING AND IMPLEMENTING A MONITORING PLAN

7.1 PROCEDURE

In order to insure that the implemented corrective and preventive actions are leading to long-term sustainable compliance, a monitoring plan needs to be developed and implemented. The monitoring phase of the laboratory audit system is designed to take advantage of all of the processes, tools, and skills developed throughout preparation, audit and data capture, reporting, corrective action, and verification phases. The long-term goal of the monitoring plan is to insure that the effort expended during the previous phases is not wasted and that the organization does not become recidivistic (e.g., return to the same old habits) with respect to CGMP compliance. This is important because the FDA frequently encounters recidivism within companies that have been subject to regulatory actions. Moreover, implementation of a monitoring plan confirms successful execution of systems-based corrective and preventive actions to previously identified gaps. The steps in implementing and executing the monitoring plan are shown Figure 7.1 and described in Table 7.1.

Establishing a CGMP Laboratory Audit System. By David M. Bliesner
Copyright © 2006 John Wiley & Sons, Inc.

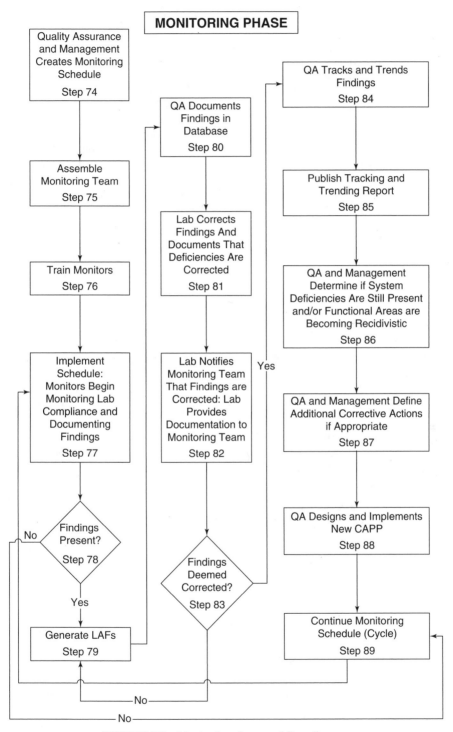

FIGURE 7.1 Monitoring phase workflow diagram.

TABLE 7.1 Explanation of Monitoring Phase Workflow Diagram Steps

Step	Description	Explanation
74	Quality assurance and management create monitoring schedule	During the monitoring phase, the quality assurance unit takes the lead role and works with management to develop and implement a monitoring schedule. This schedule is simply an extension of QA's normal audit routine. However, QA now has data to show where problems existed in the past and a better idea of where problems may occur in the future. Random sampling of previous deficiencies is recommended in developing the monitoring schedule.
75	Assemble monitoring team	The monitoring team may be composed primarily of QA personnel. However, it is recommended to include previous members of the audit, corrective action, and verification teams whenever possible. In addition, adding new team members helps to insure organizationalwide buy-in with sustainable compliance. Ownership can enhance success, especially with respect to compliance.
76	Train monitors	Monitors should be trained using a similar approach in order to train audit team members. Use as many of the tools and techniques developed previously as appropriate.
77	Implement schedule: Monitors begin monitoring lab compliance and documenting findings	Monitors use the same auditing techniques and tools as used during the Audit phase. Findings should initially be documented in audit notebooks or on checklists. Collect and compile supporting data as necessary.
78	Findings present?	If the preparation-to-verification phases were executed with due diligence, the number of deficiencies and findings should be relatively minimal. However, one should never expect the laboratory to be finding free. The nature and complexity of the laboratory control system is such that some findings and deficiencies will most likely still exist, regardless of the care and effort undertaken.
79	Generate LAFs	Findings and deficiencies are documented as before on laboratory audit forms.
80	QA documents findings in database	QA enters these findings into its existing database. This database may be the same when

TABLE 7.1 (*Continued*)

Step	Description	Explanation
		used during execution of the corrective action project plan (CAPP).
81	Lab corrects findings/and documents that deficiencies are corrected	Laboratory management is responsible for correction of the findings and submitting formal documentation that they have been corrected. This is mostly likely accomplished by simple memoranda with inclusion of supporting documentation.
		At this point, it is expected that the findings or deficiencies will be simple lapses in the previously implemented systems-based corrective and preventive actions. If the functional area manager suspects continued deficiencies in the laboratory control system subelement, they need to approach QA and senior management to develop a plan to readdress the deficiency.
82	Lab notifies monitoring team That findings are corrected: Lab provides documentation to monitoring team	Once corrected, proof of the correction, such as a memorandum, is forwarded to the monitoring team by laboratory management.
83	Findings deemed corrected?	The monitoring team determines whether the findings and deficiencies have been corrected. If they have been corrected, a verification report is completed and filed with the appropriate supporting data. If not, additional LAFs are generated and entered into the database. The generation of additional LAFs builds the case for continued systems' deficiencies if they exist.
84	QA tracks and trends findings	Using the database, QA tracks and trends findings and deficiencies. Tracking is the process of capturing the data and trending is determining whether the problems are reoccurring.
85	Publish tracking and trending report	Tracking and trending is an important part of the monitoring plan and an integral part of sustainable compliance. A report, which may be generated using the tracking database software, should be prepared and published on a regular basis. Reporting should occur no less than once a quarter, if not more frequently.

(*Continued*)

TABLE 7.1 *(Continued)*

Step	Description	Explanation
86	QA and management determine if system deficiencies are still present and/or functional areas are becoming recidivistic	A review of the tracking and trending report is done with due diligence. Care is given to determine if system deficiencies still exist and if there are signs that the corrective and preventive actions were not effective. The worst case scenario is that the functional area or organization has become recidivistic or slipped back into their old bad habits.
87	QA and management define additional corrective actions if appropriate	QA and management may not be satisfied with the laboratory's efforts to correct findings and deficiencies discovered during the monitoring phase. They may feel that further systems-based actions are needed. If this is the case, they work with the functional area managers to identify the root causes (whether new or continuing) and develop additional corrective actions to mitigate the findings and deficiencies.
88	QA designs and implements new CAPP	If additional corrective actions are defined, QA creates and implements a new corrective action project plan.
89	Continue monitoring schedule	The monitoring phase is a continuous process for insuring sustainable compliance. Organizations failing to recognize this are destined to repeat the same mistakes and will continue to generate Band-Aid solutions to systems-based problems. Companies must recognize that compliance with CGMPs through systems-based solutions is common sense and good business. A systems-based approach makes the organization more efficient and more profitable in the long run.

CHAPTER 8

A SUMMARY FOR ESTABLISHING A CGMP LABORATORY AUDIT SYSTEM

8.1 A BRIEF REVIEW OF THE GUIDE

Over the course of the preceding seven chapters, the process of establishing a CGMP laboratory audit system has been discussed in detail.

The critical features of these discussions included:

- FDA's definition of the Laboratory Control System (LCS)
- Review of some additional subsystems or subelements that may constitute the LCS and include:
  - Laboratory managerial and administrative systems (MS)
  - Laboratory documentation practices and standard operating procedures (OP)
  - Laboratory equipment qualification and calibration (LE)
  - Laboratory facilities (LF)
  - Methods validation and technology transfer (MV)
  - Laboratory computer systems (LC)
  - Laboratory investigations (LI)
- Reasons for auditing your laboratory

Establishing a CGMP Laboratory Audit System. By David M. Bliesner
Copyright © 2006 John Wiley & Sons, Inc.

- Goals of auditing your laboratory
- Phases associated with a laboratory audit including:
  - Preparation phase
  - Audit and data capture phase
  - Reporting phase
  - Corrective action phase
  - Verification phase
  - Monitoring phase

Wherever possible, Chapters 2–7 describe details of the design and implementation for each phase. Numerous tools, templates, and examples that assist in the implementation and execution of a laboratory audit system are included in all chapters and appendixes.

In addition to the nuts and bolts presented in these chapters, the end users of this guide should consider the following additional lessons.

8.2   ADDITIONAL LESSONS FOR THE END USER

8.2.1   A Proven Approach

The design and implementation of each phase described are based on real world experiences. Moreover, the approach was proven in real circumstances to be effective in upgrading the status of the laboratory control system in several different organizations. Although it may seem somewhat excessive to the uninitiated, the approach works and insures long-term or sustainable compliance with CGMPs in your laboratories.

8.2.2   Applicability to Your Facility

Since the techniques shared here are a collage of different experiences, it was valuable to choose a model laboratory to assist in the creation of this guide. The model used was a mid-sized quality operations department colocated with a pharmaceutical manufacturing facility. The model quality operations department includes one or more finished product release testing laboratories, a raw materials testing laboratory, and a stability testing laboratory. Total personnel working in the combined laboratories was assumed to be about 50 employees including bench personnel, supervisors, managers, and support personnel. The model laboratories are assumed to have significant shortcomings in their laboratory control systems and corresponding subelements. The model also assumed the laboratories would be subject to FDA regulatory action if significant upgrades to the laboratory control systems are not made.

This model may or may not reflect the character of a particular organization, and therefore, the end user may be inclined to feel that the use of the processes and techniques described in this guide are not applicable to your current situation. However, this guide was constructed with the major characteristics in mind, namely:

- *Scalable*   This audit approach described in the guide is useful regardless of the size of the facility. It works whether your organization has 10, 100, or several hundred employees. Simply scale the magnitude of the audit based on the availability of resources at your facility and to match those laboratories which constitute your quality operations.
- *Modifiable*   The tools and templates shown in the guide are designed not only to instruct but to be copied and modified. Take them and modify them a little or modify them a lot. They are meant to save you time and prevent you from having to reinvent the wheel.

8.2.3   The Value of Systems-Based Solutions

Throughout the guide, the need for systems-based solutions to corrective actions has been emphasized. So what exactly does that mean? Quite simply it means asking the following questions:

- How do we do this?
- Why do we do this?
- What can we do to make it better?

Then you have to make the decision to do it better and commit the proper resources to accomplish your goals.

The following steps describe in more detail how to identify and correct problems so they have less chance of reoccurring.

1. Map (flowchart, process diagram, storyboard, etc.) how you accomplish your work.
2. Analyze where problems exist in the workflow process.
3. Identify the root causes of the problems.
4. Eliminate tasks that don't add value or contribute to the root causes of the problem.
5. Remove the personnel from the workflow process who do not add value or contribute to the root causes of the problem.
6. Add tasks that *do* add value and mitigate the root causes of the problem.

7. If needed, purchase resources (e.g., instrumentation, computer systems, etc.) that improve the workflow process.

8. If needed, hire or redeploy existing personnel who improve the workflow process.

9. Install and qualify the material resources as necessary.

10. Train or retrain the personnel.

11. Implement the changes.

12. Supervise the changes.

13. Monitor the changes.

8.2.4   No Immunity: Every Laboratory Is a Potential Compliance Accident in the Making

After reviewing the checklists contained in the appendixes, you may be inclined to say that your laboratories are in control and therefore, it is not necessary to implement such an audit system. Be warned that no laboratory is immune to problems. It has been said that in every analytical chemistry laboratory, things have gone wrong in the past, something has just gone wrong, something is about to go wrong, or something has gone wrong and nobody has the nerve to tell you about it yet. The fewer compliance issues a laboratory has, the easier it is to implement such a system and the easier it is to keep your lab in a state of compliance. Why wait until you really have problems?

8.2.5   Audits as Learning Tools

Throughout the guide, it is emphasized that the audit approach can in many circumstances serve as an excellent learning tool. This point cannot be overemphasized. In particular, if a manager is new to the facility and really wants to get a handle on the status of compliance and a solid understanding of how the operation works, he or she should consider organizing and participating in an audit. The integration of new personnel into the monitoring team, once the audit system has been established, is also a good way to use audits as learning tools. You never really know your organization until you have scrutinized it very carefully.

8.2.6   The Linkage Between Ownership and Success

When you implement a CGMP laboratory audit system, get as many people involved within your organization as you possibly can. Corrective action owners should be the personnel who are close to the problems. Make them accountable and responsible for systems-based solutions, but give them the authority to correct the problems. Balance the ARA (accountability–authority–responsibility) triangle and set them up for success. Make this part of their employee evaluation goals. In addition, turn audits and corrective actions into challenges versus drudgery. Make people realize the more in compliance the laboratory is, the fewer problems there will be and the less stressful their workplace will become.

8.2.7 Compliance Is Good Business

In conclusion, compliance with CGMPs and specifically the implementation of an effective laboratory control system is also good business. In order to be in compliance with CGMPs, you need to have the proper organizational structure, have well-designed and well-executed systems, and clearly defined work functions. In order to have a profitable business, you need to also have proper organizational structure, well-designed and well-executed systems, and clearly defined work functions. It's all the same thing. Compliance isn't an imposition by the federal government; it's a way of doing things better and safer.

Many circumstances have shown what we call the "5-to-1 rule" when comparing laboratories who have a good command of CGMPs compared to those who don't. What is the 5-to-1 rule? The 5-to-1 rule states that a laboratory with a good command of CGMPs can accomplish the same amount of work in the same amount of time as a laboratory with five times the personnel resources. We have literally seen laboratories of 8 people easily performing the same work functions with the same workloads as laboratories of 40 people. And the only difference the 8-person lab had over the 40-person lab was the use of a systems-based approach to comply with CGMPs. Now that's good business!

APPENDIX I

EXAMPLE AUDIT CHECKLISTS: LABORATORY SUBELEMENTS

The most efficient and effectively executed audits occur when checklists are used as the foundation for inquiry. Checklists should be the core instrument for execution of any audit. An audit should not be undertaken without the creation and use of some type of checklist or similar interview instrument.

Checklists may be used in a several fashions. First, they may be used as guide to help structure interviews and documentation collection and review. After the interviews and document review, checklists can be completed to determine whether all of the appropriate issues have been covered. Second, they may also be used strictly as checklists, in a question-and-answer form, determining whether the laboratory is in compliance with the CGMP components of the subelement. Third, checklists may be handed out to in advance to interviewees for completion with confirmation of the answers by the auditors during follow-up interviews. The audit team leader should review the corresponding checklist with each subelement audit team as part of the training process. Checklists may (and should) be modified as appropriate. This may include adding, removing, or modifying questions as appropriate.

Example checklists corresponding to all seven laboratory control system subelements follow. Considerable effort has been made to insure these checklists reflect current industry thinking with respect to interpretation and

Establishing a CGMP Laboratory Audit System. By David M. Bliesner
Copyright © 2006 John Wiley & Sons, Inc.

compliance with the CGMPs as they relate to laboratory operations. Undoubt-edly, however, these checklists do not capture every facet of compliance for every laboratory within the pharmaceutical industry, nor in many circum-stances be completely applicable to individual laboratory operations. There-fore, the end user is encouraged to use these example checklists as a guide and as a backbone for development of his or her own customized and tailored documents.

LABORATORY MANAGERIAL AND ADMINISTRATIVE SYSTEMS (MS) SUBELEMENT CHECKLIST

Laboratory managerial and administrative systems are those subelements which provide the infrastructure for efficient and compliant operations of an analytical laboratory. It is sometimes difficult to directly relate items within this subelement to CGMPs. However, deficiencies within this subelement that cannot be directly attributed to the CGMPs can lead to compliance failures. For example, 21 CFR Sections 210 and 211 do not specifically require that someone be assigned the duties of training manager. However, failing to have an individual responsible for training and consequently not have an effective training program can degrade the overall quality of data generated by laboratory personnel.

Laboratory managerial and administrative systems subelement includes at least eight individual topics. A laboratory in compliance with CGMPs should address each of these topics (as applicable). These topics include: (1) Organizational structure and roles and responsibilities, (2) Training, (3) Tracking and trending-statistical quality control, (4) Complaints, (5) Laboratory purchasing and requisition, (6) Laboratory administration, (7) Laboratory chemicals, solutions, reagents and supplies, and, (8) Laboratory reference standards and solutions.

These eight topics are outlined in the following lists along with some suggestions on what items should be addressed in each topic. The list of items under each topic is fairly comprehensive but may not be totally include all those components constituting the subelement.

Step	Yes	No	Na	Laboratory Managerial and Administrative Systems Subelement (MS)
1.1				*Organizational Structure and Roles & Responsibilities*
1.1.1				Are current organization charts available and accurate?
1.1.2				Is there a policy/procedure that defines the responsibility and authority of the quality control unit?

Checklist (*Continued*)

Step	Yes	No	Na	Laboratory Managerial and Administrative Systems Subelement (MS)
1.1.3				Have the responsibilities of each functional group and supervisor been clearly defined, including testing and operational requirements, SOPs, and all other critical functions?
1.1.4				Is an organizational structure in place which is properly staffed to assure that all required testing/monitoring and support activities are performed?
1.1.5				Is the span of control and authority assigned to the quality control adequate to allow proper execution of these activities?
1.1.6				What is the ratio of supervision to analyst? (8:1 recommended)
1.1.7				Are the roles and responsibilities for each position clearly defined?
1.1.8				Are job descriptions available?
1.1.9				Are signature authority, responsibility, and accountabilities appropriate and clearly defined?
1.1.10				Do systems exist to enhance communications, understanding, and working relationships between laboratory and quality assurance personnel?
1.1.11				Does a personnel performance evaluation system exist which tracks laboratory personnel strengths and weaknesses and establishes corrective action procedures to mitigate any weaknesses?

(*Continued*)

Checklist   (*Continued*)

Step	Yes	No	Na	**Laboratory Managerial and Administrative Systems Subelement (MS)**
1.1.12				Does a master testing schedule or similar document(s) exist to insure smooth workflow and minimize laboratory personnel overcommitment?
1.1.13				Do current CVs and resumes exist for all personnel including consultants?
1.2				*Training*
1.2.1				Have the educational, training and work experience requirements for each laboratory position been clearly defined and do they reflect current standards in the industry?
1.2.2				Are training requirements clearly documented in an SOP or similar guidance document (including managers, supervisors, analysts and temporary staff)?
1.2.3				Has a training curriculum been developed for each position that clearly identifies training requirements for all required SOPs and policies, safety procedures, CGMPs as well as all other internal and external courses or programs? (database or hard copy)?
1.2.4				Is job-specific training identified and executed (e.g., laboratory analyst's qualification system)?
1.2.5				Does each employee have a training file or similar record?
1.2.6				Are the training histories for each individual employee kept current (database or hard copy)?

Checklist   (*Continued*)

Step	Yes	No	Na	Laboratory Managerial and Administrative Systems Subelement (MS)
1.2.7				Have all laboratory personnel been properly trained?
1.2.8				Is this training documented and are the documents available for review?
1.2.9				Is SOP training conducted in a fashion other than read-and-understand?
1.2.10				Are metrics used to determine effectiveness of SOP training?
1.2.11				Is there an analyst qualification program in place?
1.2.12				Has an individual been designated as the training coordinator or manager?
1.2.13				Is there evidence of management support for training and training programs?
1.2.14				Does a formal training budget exist?
1.2.15				Do vendor training records exist?
1.2.16				Is there evidence that newly hired employees are evaluated for skill level and competency?
1.2.17				Is there a formal training schedule in place and is it being executed?

(*Continued*)

Checklist *(Continued)*

Step	Yes	No	Na	Laboratory Managerial and Administrative Systems Subelement (MS)
1.2.18				Is there evidence of employee retraining and requalification?
1.2.19				Is there a positive attitude with respect to training on the part of management and laboratory personnel?
1.3				*Tracking and Trending—Statistical Quality Control*
1.3.1				Is tracking and trending (e.g., control charting) of data performed with the following points to be considered:
1.3.1.1				For finished product testing?
1.3.1.2				For raw material testing?
1.3.1.3				Procedures delineated in an SOP?
1.3.1.4				Appropriate periodic evaluation of control chart data included annual product reviews?
1.3.1.5				Appropriateness of response to evaluations (e.g., excursions or trends are formally addressed as per SOP?)
1.3.1.6				For finished product-stability data (release and retain samples?)
1.4				*Complaints*
1.4.1				Is there a formal product complaint handling system in place, including some system of complaint monitoring and statistical review with the following points to be considered:

Checklist (*Continued*)

Step	Yes	No	Na	Laboratory Managerial and Administrative Systems Subelement (MS)
1.4.1.1				Procedures delineated in an SOP?
1.4.1.2				Site procedures consistent with standard industry practice?
1.4.1.3				Quality control laboratories are involved and provide testing as appropriate?
1.5				*Laboratory Purchasing and Requisition*
1.5.1				Are levels of approval defined and appropriate for purchasing and requisitions for the laboratory?
1.5.2				Are items purchased from qualified vendors (e.g., raw materials, reagents, standards, instruments, etc.)?
1.5.3				Does the laboratory have a defined yearly budget? Is the budget derived from existing data and projected workloads?
1.5.4				Does the laboratory have a defined capital budgeting process? Is the budget derived from existing data and projected needs?
1.5.5				Does the laboratory have a defined process for requesting and receiving approval to hire new personnel?
1.5.6				Does the laboratory have specific personnel assigned to purchasing and budgeting activities?

(*Continued*)

Checklist (*Continued*)

Step	Yes	No	Na	**Laboratory Managerial and Administrative Systems Subelement (MS)**
1.5.7				Do these personnel have the proper skills and training to perform planning, budgeting, and purchasing activities?
1.6				*Laboratory Administration*
1.6.1				Does the laboratory have a general administrator, office manager, or similar administrative support person?
1.6.2				Is someone assigned the responsibility for:
1.6.2.1				Managing work orders for office and laboratory repairs?
1.6.2.2				Budget preparation, oversight, and review?
1.6.2.3				Overseeing stockroom operations?
1.6.2.4				Supervising requisition of administrative and operational supplies?
1.6.2.5				Managing external contractors and contracts?
1.6.2.6				Managing/enhancing communication between the laboratory and internal/external customers?
1.6.2.7				Supervising laboratory safety training and compliance?
1.6.2.8				Monitoring overtime billing?

Checklist   (*Continued*)

Step	Yes	No	Na	**Laboratory Managerial and Administrative Systems Subelement (MS)**
1.6.2.9				Capital equipment purchases?
1.6.2.10				Supervises administrative service personnel such as maintenance and calibration, laboratory computers, documentation section, and training section?
1.6.3				Are there laboratory corrective and preventive action committees or similar groups in place?
1.6.4				Are the laboratory corrective and preventive action committees' meeting minutes published and proposed actions communicated to all appropriate personnel?
1.6.5				Are data reported in all types of reports easily traceable to raw data?
1.7				*Laboratory Chemicals, Solutions, Reagents, and Supplies*
1.7.1				Is there a written procedure for receipt and storage of chemicals and reagents?
1.7.2				Are chemicals and reagents properly labeled with date of receipt, date opened, and expiration and retest dates?
1.7.3				Are laboratory prepared reagents and solutions properly identified? (e.g., chemical name or symbol, concentration, date of preparation, initials of the analyst who prepared it, and expiration date)?

(*Continued*)

Checklist   (*Continued*)

Step	Yes	No	Na	Laboratory Managerial and Administrative Systems Subelement (MS)
1.7.4				Are records available to document preparation and standardization of volumetric solutions?
1.7.5				Is the frequency of standardization of various reagents described? Is it consistent with the current USP?
1.7.6				Are standardized reagents properly stored to assure integrity?
1.7.7				Are there procedures that describe the preparation of laboratory reagents and cultures?
1.7.8				Are there procedures that describe the maintenance of laboratory reagents and cultures?
1.8				*Laboratory Reference Standards and Solutions*
1.8.1				Is there a written procedure for ordering and receipt of compendial reference standards and noncompendial reference standards?
1.8.2				Are the primary standards of the current lot listed in the USP, EP, or JP?
1.8.3				Is the receipt of the standards logged?
1.8.4				Are all standards labeled with name, source, lot number, and expiration date?

Checklist (*Continued*)

Step	Yes	No	Na	Laboratory Managerial and Administrative Systems Subelement (MS)
1.8.5				Do written procedures include provisions for prevention of contamination of primary standards?
1.8.6				Are standards stored in a secured area under environmentally controlled and monitored conditions?
1.8.7				Are procedures for assuring standard integrity available?
1.8.8				Are working or house standards checked against primary standards at appropriate intervals?
1.8.9				Are stock solutions appropriately identified and are their expirations based on their true stability?
1.8.10				Do procedures exist for the certification and use of noncommercially available reference standards?
1.8.11				Do certificates of analysis exist for all reference standards and are these certificates stored as controlled documents?
1.8.12				Have provisions been made for handling controlled substance reference standards?
1.8.13				Does the reference standard SOP address (map) proper handling of controlled substance reference standards?
9.0				*Additional Items*

LABORATORY DOCUMENTATION PRACTICES AND STANDARD OPERATING PROCEDURES (OP) SUBELEMENT CHECKLIST

Laboratory documentation practices and standard operating procedures form a critical component of the overall Laboratory Control System. Much of the essence of CGMPs involves the generation, review, approval, revision and control of procedures and data. In compliance parlance, "If it isn't written down, it didn't happen."

Laboratory documentation practices and standard operating procedures subelement includes at least nine individual topics. A laboratory, which is in compliance with CGMPs, should address each of these topics (as applicable). These topics include: (1) SOPs—General, (2) SOPs—specific topics, (3) Laboratory test procedures, (4) Laboratory data and results, (5) Security of data, (6) Distribution of results, (7) Chromatography, (8) In-process testing, and (9) Assignment of retest/expiry dates.

The following text lists nine topics along with some suggestions on what items should be addressed in each topic. The list of items under each topic is fairly comprehensive but may not be totally inclusive of all those components constituting the subelement.

Step	Yes	No	Na	Laboratory Documentation Practices and Standard Operating Procedures Subelement (OP)
2.1				*SOPs—General*
2.1.1				Is there a comprehensive and current laboratory document control system in place?
2.1.2				Is there a list of all approved SOPs?
2.1.3				Are the SOPs current, clearly written, and accessible to all appropriate personnel?
2.1.4				Is there a system for periodic review of all SOPs to insure that they are consistent with current company and industry practices?

Checklist   (*Continued*)

Step	Yes	No	Na	**Laboratory Documentation Practices and Standard Operating Procedures Subelement (OP)**
2.1.5				Is there an SOP governing the SOP or document control system (e.g., an SOP on SOPs)?
2.1.6				Is there a system for controlling the issuance and revision of all SOPs?
2.1.7				Do policies and manuals exist which supplement the SOPs? (e.g., Level I and Level II guidance documents)
2.1.8				Are the guidance documents readily accessible to all laboratory personnel?
2.1.9				Are all SOPs reviewed and updated at least every year?
2.2				*SOPs—Specific Procedures*
				Are there specific standard operating procedures covering:
2.2.1				Change control?
2.2.2				Record review and approval requirements?
2.2.3				Description of the requirements for performing data review?
2.2.4				Laboratory records (e.g., data recorded in bound notebooks or controlled worksheets)?
2.2.5				Dissolution testing?

(*Continued*)

Checklist (*Continued*)

Step	Yes	No	Na	**Laboratory Documentation Practices and Standard Operating Procedures Subelement (OP)**
2.2.6				Sample receipt, documentation, handling, storage, and control?
2.2.7				Laboratory investigations, deviations and handling of OOS results?
2.2.8				Policy for identification and reporting of new impurities and/or higher levels of previously known impurities?
2.2.9				Validation and transfer of analytical methods?
2.2.10				Laboratory computer validation?
2.2.11				Preparation, labeling, identification, expiration-dating, and storage of chemicals, reagents, and solutions?
2.2.12				Equipment/Instrument use, PM, calibration and qualification?
2.2.13				Appropriate labeling of out of use lab equipment?
2.2.14				Glassware washing (mechanical and manual)?
2.2.15				Practices regarding housekeeping, safety glasses, eating in labs, smoking, etc.?
2.2.16				Hazardous waste disposal?

Checklist (*Continued*)

Step	Yes	No	Na	**Laboratory Documentation Practices and Standard Operating Procedures Subelement (OP)**
2.2.17				Handling and disposing laboratory samples?
2.2.18				Is there an SOP for defining appropriate sampling plans for all quality control laboratory testing?
2.2.19				Are samples randomly chosen and are they representative of all portions of the lot(s)?
2.2.20				Is there an SOP for the receipt, documentation, handling, storage and distribution of laboratory samples?
2.2.21				Is a sample logbook maintained?
2.2.22				Are samples tracked? Is sample disposition included in tracking?
2.2.23				Are storage areas maintained for sample disposition (e.g., retention and destruction)?
2.2.24				Is there a master signature list, including persons' names, initials, and responsibilities?
2.2.25				Are samples labeled appropriately to include: sample description, source, quantity, date sampled, date sample received for testing?
2.2.26				Are sample storage areas properly identified and maintained?

(*Continued*)

Checklist (*Continued*)

Step	Yes	No	Na	Laboratory Documentation Practices and Standard Operating Procedures Subelement (OP)
2.2.27				Is there a backlog with respect to sample analysis, and if so is the backlog great enough to have an impact on testing results and product release?
2.3				*Laboratory Test Procedures*
2.3.1				Is there an index listing the testing documents and specifications?
2.3.2				Is this list updated and properly controlled?
2.3.3				Do the test procedures include sufficient instructions on how to conduct the testing and operate the specific lab instruments?
2.3.4				Is there a system for controlling the issuance and revision of all testing related records?
2.3.5				Are laboratory test procedures and specifications approved by the quality assurance unit?
2.3.6				Is there a written change control procedure that describes how analytical method changes are made?
2.3.7				Do specifications for compendial products meet compendial requirements?
2.3.8				Is there a procedure for determining in-house or tighter release specifications?

Checklist (*Continued*)

Step	Yes	No	Na	**Laboratory Documentation Practices and Standard Operating Procedures Subelement (OP)**
2.3.9				Are raw materials released by using validated analytical methods?
2.4				*Laboratory Data and Results*
2.4.1				Is data documented in bound prenumbered logbooks, notebooks, or other data storage and acquisition systems?
2.4.2				Is the use of scrap paper, Post-its, or similar uncontrollable paper specifically prohibited in the laboratory?
2.4.3				Are corrections to data entry errors covered in an SOP?
2.4.4				Are all handwritten documents/data recorded in permanent ink?
2.4.5				If bound notebooks are used, is there a system for control, issuance, use, and return?
2.4.6				Is each notebook page dated and signed by the analyst and second-reviewing authorized individual and performed in a timely manner?
2.4.7				Are all entries checked and approved for completeness of sample identity, proper use of reagents, standards, and execution of experimental procedures?

(*Continued*)

Checklist (*Continued*)

Step	Yes	No	Na	Laboratory Documentation Practices and Standard Operating Procedures Subelement (OP)
2.4.8				Are all outputs, such as chromatograms, spectra, etc., appropriately identified, stored, referenced to notebooks, and readily retrievable?
2.4.9				Are all data, calculations, and results verified by second qualified individual for accuracy, completeness, and compliance with specifications?
2.4.10				Are the qualifications of persons conducting data review documented and acceptable?
2.4.11				Is data review documented by full signature and date?
2.4.12				Are data review and approval activities captured in LIMS?
2.4.13				Are data reviewed by the laboratory and approved by quality assurance before distribution? Is this process defined by an SOP?
2.4.14				Is there a system in place to assure that the method being used in the laboratory is the current approved-version method for testing?
2.4.15				Is the method used for analysis recorded onto the laboratory testing results sheet or recorded into LIMS?
2.4.16				Is there a written SOP for handling of reintegration of HPLC/GC data?

Checklist (*Continued*)

Step	Yes	No	Na	**Laboratory Documentation Practices and Standard Operating Procedures Subelement (OP)**
2.4.17				If a computer software package is used to calculate results, is there an example calculation included with the reported results verifying proper algorithm execution?
2.4.18				Are all correction factors used in calculations expressed with the proper units?
2.4.19				Are raw data defined in an SOP?
2.5				*Security of Data*
2.5.1				Is access to stored data limited?
2.5.2				Are data protected from fire, water, and other environmental hazards?
2.5.3				Are data readily retrievable?
2.5.4				Is the length of time that documents need to be retained defined in writing and does this time frame meet current industry practice?
2.5.5				Are backup copies of data stored at an offsite location?
2.5.6				Is there a disaster recovery plan and has it been rehearsed?

(*Continued*)

Checklist (*Continued*)

Step	Yes	No	Na	Laboratory Documentation Practices and Standard Operating Procedures Subelement (OP)
2.6				*Distribution of Results*
2.6.1				Are results entered into a LIMS or other electronic management system?
2.6.2				Is LIMS used to compare testing results to the specifications for determining pass/fail decisions?
2.6.3				Is LIMS used to provide data for use in annual product reviews?
2.6.4				If data are transcribed from notebooks and reports, is transcription accuracy being verified?
2.7				*Chromatography*
2.7.1				Is system suitability performed routinely? (e.g., HPLC methods, GC methods, etc.)
2.7.2				Are the system suitability parameters determined from validation data?
2.7.3				Must system suitability parameters be calculated and met prior to executing a quality control sample analysis?
2.7.4				Is there an SOP that delineates a standard sample queue (i.e., interspersing of standards and samples during the course of the chromatographic run)?

Checklist   (*Continued*)

Step	Yes	No	Na	**Laboratory Documentation Practices and Standard Operating Procedures Subelement (OP)**
2.8				*In-Process Testing*
2.8.1				Are validated methods used for in-process testing?
2.8.2				Do formal sampling plans exist for all in-process testing?
2.8.3				Are there established acceptance criteria/ranges/specifications for in-process tests?
2.8.4				Is sample integrity preserved during handling and storage of in-process samples?
2.8.5				Are all the necessary industry standard in-process tests conducted (i.e., blend uniformity, weight variation, hardness, thickness, etc.)?
2.9				*Assignment of Retest/Expiry Dates*
2.9.1				Are expiration/retest dates based on available in-house storage (stability) data or otherwise justified?
2.9.2				Does an SOP exist which differentiates between retest and expiry dating?
2.9.3				Is there a limit for extending the retest life of the substance evaluated?

(*Continued*)

Checklist (*Continued*)

Step	Yes	No	Na	**Laboratory Documentation Practices and Standard Operating Procedures Subelement (OP)**
2.9.4				Are there procedures for handling/storage of reference standards and reference cultures?
9.0				*Additional Items*

LABORATORY EQUIPMENT QUALIFICATION AND CALIBRATION (LE) SUBELEMENT CHECKLIST

Since the vast majority of data in a modern analytical laboratory are acquired via instrumentation, it is essential that these instruments have been properly installed, qualified, and are kept in proper working order. The laboratory equipment qualification and calibration subelement addresses these issues.

Laboratory equipment qualification and calibration include at least three individual topics. A laboratory, which is in compliance with CGMPs, should address each of these topics (as applicable). These topics include: (1) Laboratory equipment procedures—General, (2) Laboratory equipment procedures—Specific, and (3) Laboratory equipment procedures—Computer controlled.

The following text lists these three topics along with some suggestions on what items should be addressed in each topic. The list of items under each topic is fairly comprehensive but may not be totally inclusive of all those components constituting the subelement.

Step	Yes	No	Na	Laboratory Equipment Qualification and Calibration Subelement (LE)
3.1				*Laboratory Equipment Procedures—General*
3.1.1				Are the laboratories equipped with all of the necessary instruments for the analytical testing to be performed?
3.1.2				Is there a written qualification, calibration, and preventive maintenance program in place described in an SOP?
3.1.3				Is there a master equipment list available and is it properly maintained?
3.1.4				Is there a master equipment maintenance and calibration schedule?
3.1.5				Are specific personnel assigned to maintain the master equipment list and equipment maintenance and calibration schedule?

(*Continued*)

Checklist *(Continued)*

Step	Yes	No	Na	Laboratory Equipment Qualification and Calibration Subelement (LE)
3.1.6				Are calibration/service vendors qualified and are their training records available?
3.1.7				Is there a validation master plan indicating timing and responsibility for equipment qualification and requalification?
3.1.8				Are there triggers (e.g., specific criteria and/or circumstances) for requalification and are they integrated into the master plan?
3.1.9				Does each piece of equipment have a logbook or file documenting instrument maintenance, calibration, and repair histories?
3.1.10				Does each piece of equipment have a usage log?
3.1.11				Do labels on the instruments identify the person who performed the calibration, date of calibration, and due date of next calibration?
3.1.12				Are IQ, OQ, and PQ requirements, which clearly define performance testing acceptance criteria for each piece of equipment, clearly defined in an approved protocol?
3.1.13				Have IQ/OQ/PQ been performed and properly documented for all equipment?
3.1.14				Are IQ, OQ, and PQ performed during initial system installation?

Checklist (*Continued*)

Step	Yes	No	Na	Laboratory Equipment Qualification and Calibration Subelement (LE)
3.1.15				Is calibration and preventive maintenance (PM), including tolerances, based on the manufacturer's recommendations?
3.1.16				Are cleaning procedures included in the SOPs for instrument qualification or calibration?
3.1.17				Is there an SOP that requires that instruments failing calibration are removed from service?
3.1.18				Does the SOP provide directions for investigations when instrumentation is found to be out-of-calibration, yet still is being used to generate CGMP data?
3.1.19				Are the qualification specifications for equipment based on sound scientific principals?
3.1.20				Are calibration/metrology records secure (locked up)?
3.2				*Laboratory Equipment Procedures—Specific*
3.2.1				Are balances checked for corner loading?
3.2.2				Are pH meters standardized and calibrated?
3.2.3				Are balances calibrated at both upper and lower weighing capability using NIST-traceable standards?

(*Continued*)

Checklist   (*Continued*)

Step	Yes	No	Na	Laboratory Equipment Qualification and Calibration Subelement (LE)
3.2.4				Are calibration correction factors attached to thermometers (if appropriate)?
3.2.5				Are pH meters calibrated at two points, not differing by more than 4 pH units?
3.2.6				Are UV spectrophotometers calibrated for wavelength and photometric accuracy?
3.2.7				Is there a written procedure for set-up, calibration, and operation of dissolution baths?
3.2.8				Are there written procedures for calibration and maintenance of HPLCs, including pumps, auto sampler reproducibility, and wavelength accuracy?
3.2.9				Do HPLC PMs include seal and valve changes?
3.3				*Laboratory Equipment Procedures—Computer Controlled*
3.3.1				Is there a written policy or procedures for validation of computer-interfaced instrumentation?
3.3.2				Have the computer systems/software, which control instruments, been validated?

Checklist (*Continued*)

Step	Yes	No	Na	Laboratory Equipment Qualification and Calibration Subelement (LE)
3.3.3				Is there a system in place to install new software versions and to decide what scope of revalidation must be done during software upgrades?
3.3.4				Is computer software validation a part of the OQ process?
3.3.5				Is computer software validation documented?
3.3.6				Are there systems in place to prevent installation of unauthorized software?
3.3.7				Are integrators validated using a validated signal generator?
3.4				*Additional Items*

LABORATORY FACILITIES (LF) SUBELEMENT CHECKLIST

As with a manufacturing area, it is essential that the laboratory be properly designed and equipped. This insures that the data generated accurately reflect the true composition of the sample from which they came. In addition, it insures a safe working environment for laboratory personnel.

The laboratory facilities subelement includes three individual topics. A laboratory, which is in compliance with CGMPs, should at least address each of these topics (as applicable). These topics include: (1) Laboratory facilities—General, (2) Safety and environmental concerns, and (3) Laboratory glassware.

The following text lists these three topics along with some suggestions on what items should be addressed in each topic. The list of items under each topic is fairly comprehensive but may not be totally inclusive of all those components constituting the subelement.

Step	Yes	No	Na	Laboratory Facilities Subelement (LF)
4.1				*Laboratory Facilities—General*
4.1.1				Is the physical construction of the laboratory areas adequate for testing and all routine activities with respect to:
4.1.1.1				Size?
4.1.1.2				Layout and design for optimum workflow and efficiency?
4.1.1.3				Sample receipt?
4.1.1.4				Appropriate tables for balances/instruments, etc.?
4.1.1.5				Location of SOPs, location of methods, specification sheets, and safety data sheets?

Checklist   (*Continued*)

Step	Yes	No	Na	Laboratory Facilities Subelement (LF)
4.1.1.6				Location of reagents, standards, solutions, solvents?
4.1.1.7				Data entry, recording, writing areas?
4.1.1.8				Sample storage and retention?
4.1.1.9				Refrigeration?
4.1.2				Are proper systems in place to minimize cross-contamination during sample preparation and laboratory testing?
4.1.3				Are all controlled temperature/humidity storage areas, incubators, etc. monitored to assure that proper conditions are maintained?
4.1.4				Have temperature monitoring systems and equipment been properly validated?
4.1.5				Have humidity monitoring systems and equipment been properly validated?
4.1.6				Are controlled substances properly segregated and controlled according to DEA requirements?
4.1.7				Is the environment/temperature control correct for the areas for different types of equipment?

(*Continued*)

Checklist (*Continued*)

Step	Yes	No	Na	Laboratory Facilities Subelement (LF)
4.1.8				Are radioisotopes (if used) in proper storage areas?
4.1.9				Are uninterruptible power supply units (UPS) used to minimize the impact of power surges and outages?
4.1.10				If a UPS exists, has it been qualified and has an SOP describing its use been written?
4.1.11				Are piping systems specified at the correct pressure and have they been cleaned, flushed, leak-tested passivated (sanitary use), and certified?
4.1.12				Have appropriate hood air filters been installed and smoke tests performed?
4.1.13				Have facility monitoring systems been validated?
4.1.14				Have purified water systems been validated?
4.1.15				Is there an SOP for lab or engineering maintenance of the air system?
4.1.16				Are there appropriate controls available/specifications for handling particulate matter and microbes?
4.1.17				Is there control of HVAC in instrument rooms, balance area, etc.?

Checklist (*Continued*)

Step	Yes	No	Na	**Laboratory Facilities Subelement (LF)**
4.1.18				Does the hood exhaust into an appropriate location (e.g., a sufficient distance from the air intake system as to not interfere with lab operations)?
4.1.19				Is the HVAC for stability/retention samples adequately monitored?
4.1.20				Has the laboratory water system been properly qualified?
4.1.21				Is there routine monitoring to assure adequate quality of the water?
4.1.22				Are the types of water required for the lab identified and are specifications properly set?
4.1.23				Is lab water part of the site water monitoring program and is testing frequency established and appropriate?
4.1.24				Are there action plans for following water related out of specification results or deviations?
4.1.25				Is there a schedule for servicing the laboratory water system?
4.2				*Safety and Environmental Concerns*
4.2.1				Is there a safety and environmental SOP?
4.2.2				Is there testing of hoods?

(*Continued*)

Checklist   (*Continued*)

Step	Yes	No	Na	Laboratory Facilities Subelement (LF)
4.2.3				Are hazardous materials identified and segregated?
4.2.4				Are safety data sheets available and are they kept current?
4.2.5				Is there a program that describes handling and disposal of biohazardous waste?
4.2.6				Are toxic/dangerous chemicals handled and disposed of according to local, state, and federal regulations?
4.2.7				Are employees trained to handle toxic/dangerous materials?
4.3				*Laboratory Glassware*
4.3.1				Has glassware washing/cleaning been validated to include demonstrating the removal of chemical residue?
4.3.2				Is there a PM program for mechanical glassware washers?
4.3.3				Is there a program in place to check for defective and worn glassware?
9.0				*Additional Items*

METHODS VALIDATION AND TECHNOLOGY TRANSFER SUBELEMENT (MV)

Methods validation is the process of demonstrating that analytical procedures are suitable for their intended use. The methods validation process for analytical procedures begins with the planned and systematic collection validation data to support analytical procedures. In addition, the methods validation process is considered to be complete once the methods have been successfully transferred from the originating to the receiving or end user laboratory.

The methods validation and technology transfer subelement includes at least three individual topics. A laboratory, which is in compliance with CGMPs, should at least address each of these topics. These topics include: (1) Validation of analytical methods—General, (2) Cleaning methods validation, and (3) Procedures for methods transfer.

The following text lists these three topics along with some suggestions on what items should be addressed in each topic. The list of items under each topic is fairly comprehensive but may not be totally inclusive of all those components constituting the subelement.

Step	Yes	No	Na	Methods Validation and Technology Transfer Subelement (MV)
5.1				*Validation of Analytical Methods—General*
5.1.1				Is there a general SOP for methods validation?
5.1.2				Are the requirements in the SOP consistent with FDA/ICH/USP guidelines?
5.1.3				Are methods validation protocols used for each validation?
5.1.4				Are all of the appropriate elements of validation included in the methods validation protocol such as:
5.1.4.1				Accuracy?

(*Continued*)

Checklist   (*Continued*)

Step	Yes	No	Na	Methods Validation and Technology Transfer Subelement (MV)
5.1.4.2				Precision (repeatability and intermediate precision)?
5.1.4.3				Selectivity/Specificity?
5.1.4.4				Robustness?
5.1.4.5				Linearity?
5.1.4.6				Range?
5.1.4.7				Limit of detection?
5.1.4.8				Limit of quantitation?
5.1.4.9				System suitability?
5.1.4.10				Sample solution stability?
5.1.4.11				Completeness of extraction (from dosage forms)?
5.1.4.12				Filter retention studies?
5.1.4.13				Forced degradation studies?

Checklist (*Continued*)

Step	Yes	No	Na	**Methods Validation and Technology Transfer Subelement (MV)**
5.1.5				Is there an SOP or section within an SOP that discusses the requirements for the verification of compendial methods?
5.1.6				Have compendial methods been shown to work under the actual conditions of use?
5.1.7				Have impurity methods been properly validated? Are they stability indicating and selective or specific?
5.1.8				Are method validation reports published following validation of an analytical method?
5.1.9				Can results presented in method validation reports be tracked back to raw data?
5.2				*Cleaning Methods Validation*
5.2.1				Is there an ongoing cleaning validation program in place?
5.2.2				Is cleaning validation conducted according to a master plan and schedule?
5.2.3				Are there validated sampling procedures developed to support cleaning validation studies?
5.2.4				Are there written acceptance criteria?

(*Continued*)

Checklist   (*Continued*)

Step	Yes	No	Na	**Methods Validation and Technology Transfer Subelement (MV)**
5.2.5				Were recovery studies conducted?
5.2.6				Are there validated analytical methods developed to support cleaning validation studies?
5.2.7				Does the cleaning validation testing include analysis of area swap samples?
5.3				*Procedures for Methods Transfer*
5.3.1				Are there site policies and procedures governing transfer of methods, and do these documents include defining acceptance criteria?
5.3.2				Are the responsibilities for who is in charge clearly defined (e.g. receiving laboratory or originating laboratory)?
5.3.3				Is there a requirement to form an analytical technology transfer team or similar organization to transfer methods?
5.3.4				Is a formal receiving laboratory readiness assessment performed to include evaluation of equipment, facilities, and personnel?
5.3.5				Is the receiving laboratory given the opportunity to review the analytical methodology prior to the transfer process?

Checklist (*Continued*)

Step	Yes	No	Na	**Methods Validation and Technology Transfer Subelement (MV)**
5.3.6				Is an analytical technology transfer qualification strategy document generated and used to guide the transfer of methods?
5.3.7				Are personnel in the receiving laboratory given the opportunity to use the method under the supervision of the originating personnel subject-matter experts prior to formal transfer?
5.3.8				Is an analytical technology transfer protocol, with predetermined acceptance criteria, used to formally transfer the methodology?
5.3.9				Is a reviewed and approved technology transfer report generated upon successful execution of the transfer protocol?
5.3.10				Can results presented in the cleaning methods validation reports be tracked back to raw data?
9.0				*Additional Items*

LABORATORY COMPUTER SYSTEMS (LC) SUBELEMENT

In a modern analytical laboratory, nearly every instrument is either controlled by or has data residing and processed by a computer. Subsequently, it is critical that reliability of these systems and their associated software be evaluated.

Laboratory computer systems includes at least seven individual topics. A laboratory, which is in compliance with CGMPs, should address each of these topics (as applicable). These topics include: (1) Laboratory computer systems—General, (2) Centralized and network-attached data systems, (3) Stand-alone data systems, (4) SOPs and records, (5) Laboratory information management systems (LIMS), (6) Spreadsheets, and (7) Other systems.

The following text lists these seven topics along with some suggestions on what items should be addressed in each topic. The list of items under each topic is fairly comprehensive but may not be totally inclusive of all those components constituting the subelement.

Step	Yes	No	Na	**Laboratory Computer Systems Subelement (LC)**
6.1				*Laboratory Computer Systems—General*
6.1.1				Is there an inventory of laboratory data systems (both centralized and benchtop systems) including any associated servers?
6.1.2				Is a disaster recovery plan in place addressing laboratory data systems?
6.1.2.1				Has it been tested?
6.1.2.2				Is documentation for both the plan and any tests available for review?
6.1.3				Have all GMP laboratory data systems and associated file servers etc., been validated in accordance with applicable standards and procedures?

Checklist (*Continued*)

Step	Yes	No	Na	**Laboratory Computer Systems Subelement (LC)**
6.1.4				Have all GMP laboratory data systems and associated file servers, etc. been reviewed for compliance with 21CFR Part 11?
6.1.4.1				Is there a remediation/replacement plan for those which were found to be deficient?
6.1.5				Are GMP workstations clearly labeled as such?
6.1.6				Is access to laboratory workstations provided through operating systems, having individual user accounts?
6.1.6.1				Are passwords properly controlled and protected by end users?
6.1.6.2				Are passwords properly managed with assignment of permissions and privileges?
6.1.7				Are workstations configured with password-controlled screensavers or other automatic locking mechanism?
6.1.8				Are data structures for laboratory systems documented?
6.1.8.1				Does this include defining folders/directories on local hard drives and file servers where data are stored?
6.1.8.2				Do system SOPs specify these folders/directories for data storage as appropriate?

(*Continued*)

Checklist (*Continued*)

Step	Yes	No	Na	Laboratory Computer Systems Subelement (LC)
6.1.9				Are procedures in place to govern data naming conventions for projects, analyses, products, etc.?
6.1.10				Is data archiving performed on one or more laboratory data systems?
6.1.10.1				Is the distinction between archiving and backing up data understood by laboratory staff?
6.1.10.2				Is archived data subject to formalized retention standards?
6.1.10.3				Has the dearchival process ever been tested?
6.2				*Centralized and Network-Attached Data Systems*
6.2.1				Are back-ups of data files made regularly?
6.2.1.1				Is it governed by SOP?
6.2.1.2				Is there an individual who is assigned this responsibility?
6.2.1.3				Is the back-up media securely stored, with inventory data that is current and accurate?
6.2.2				Are there documented service-level agreements with the central support organization defining mutual expectations and responsibilities?

Checklist (*Continued*)

Step	Yes	No	Na	**Laboratory Computer Systems Subelement (LC)**
6.2.3				If back-ups are performed by an automatic procedure provided by a central support organization is clear documentation in place to define expected data structures and problem alerting/resolution procedures?
6.2.4				Are data restores governed by an SOP?
6.2.5				Are the restores documented as required by SOP?
6.2.6				Does the central support organization have server/ application SOPs, which address data management, incident management, performance monitoring, etc., specific to the laboratory?
6.3				*Stand-Alone Data Systems*
6.3.1				Are back-ups performed either automatically (scheduled through a system procedure) or manually as defined by SOPs?
6.3.2				Is there a procedure in place to determine whether back-up media have been modified?
6.3.3				Are back-up media identified and securely stored?
6.3.4				Are the restores documented as required by SOP?
6.4				*SOPs and Records*
6.4.1				Do SOPs exist for validation of laboratory data systems?

(*Continued*)

Checklist (*Continued*)

Step	Yes	No	Na	**Laboratory Computer Systems Subelement (LC)**
6.4.2				Has appropriate laboratory staff been trained on the laboratory data system SOPs?
6.4.3				Are training records for personnel trained on the SOPs available for review?
6.4.4				Do SOPs exist for operation of CGMP laboratory data systems?
6.4.5				Has appropriate laboratory staff been trained on the CGMP laboratory data system operation SOPs?
6.4.6				Are training records for personnel trained on the SOPs available for review?
6.5				*Laboratory Information Management System (LIMS)*
6.5.1				If a LIMS System is used for data management (either custom or commercial system) has it been validated?
6.5.2				Has the validation been properly documented and the data available for review?
6.5.3				Is the system operated and maintained by laboratory staff or a systems support organization?
6.5.4				Is the system maintained under change control delineated in an SOP?

Checklist (*Continued*)

Step	Yes	No	Na	**Laboratory Computer Systems Subelement (LC)**
6.5.5				Have lab personnel been trained on LIMS data entry and are training records available for review?
6.5.6				Are there automated data entry and extraction points in the LIMS? Have the links been validated for:
6.5.6.1				Bar code readers?
6.5.6.2				GC data stations
6.5.6.3				HPLC data stations?
6.5.6.4				Balances?
6.5.6.5				Spreadsheets
6.5.6.6				Other components?
6.5.7				Is there a definition of what is considered raw data?
6.5.8				Can data entries be made by only authorized, password-protected individuals?
6.5.9				Are electronic signatures used in the LIMS system and are they in compliance with 21CFR Part 11?

(*Continued*)

Checklist (*Continued*)

Step	Yes	No	Na	**Laboratory Computer Systems Subelement (LC)**
6.6				*Spreadsheets*
6.6.1				How are spreadsheets used in the laboratory and if so, do they reside on the following:
6.6.1.1				Networks?
6.6.1.2				Workstation hard drives?
6.6.2				Are procedures in place governing the development/ modification of spreadsheets for GMP use?
6.6.3				Are the spreadsheets in use unalterable and:
6.6.3.1				Qualified/validated?
6.6.3.2				Are procedures in place to ensure that correct versions of spreadsheets are being used?
6.6.4				Are there procedures in place for archiving and retrieving retired versions of spreadsheets?
6.7				*Other Systems*
6.7.1				Is an electronic document management system used and if so, is it validated?
6.7.2				Is an electronic training-record database used and if so is it validated?

Checklist (*Continued*)

Step	Yes	No	Na	**Laboratory Computer Systems Subelement (LC)**
6.7.3				Are laboratory periodic maintenance and calibration activities scheduled and recorded by a maintenance management system and if so is it validated?
9.0				*Additional Items*

LABORATORY INVESTIGATIONS (LI) SUBELEMENT

CGMPs exist to not only prevent problems from occurring, but to mitigate and address deficiencies when they do occur. The laboratory investigations subelement is designed to systematically identify the root cause of failures, determine how to correct the problems, and to prevent them from reoccurring in the future.

Laboratory investigations includes at least three individual topics. A laboratory, which is in compliance with CGMPs, should address each of these topics (as applicable). These topics should include: (1) Laboratory investigations—General, (2) Laboratory investigations—Execution, and (3) Laboratory investigations—Documentation.

The following text lists these three topics along with some suggestions on what items should be addressed in each topic. The list of items under each topic is fairly comprehensive but may not be totally inclusive of all those components constituting the subelement.

Step	Yes	No	Na	**Laboratory Investigations Subelement (LI)**
7.1				*Laboratory Investigations—General*
7.1.1				Is there an SOP that defines in detail when and how laboratory investigations are conducted?
7.1.2				Does the SOP distinguish between out of specification (OOS), deviations, or atypical events?
7.1.3				Does the SOP have a flowchart and/or decision tree to assist in performing laboratory investigations?
7.1.4				Does the SOP have a checklist to assist in determining the root cause of the deviation or failure?

Checklist (*Continued*)

Step	Yes	No	Na	**Laboratory Investigations Subelement (LI)**
7.1.5				Does the checklist include review of specifics such as:
7.1.5.1				Use of proper test methods and proper execution of the test method?
7.1.5.2				Performance of proper calculations without mathematical error?
7.1.5.3				Use of proper reagents, standards, and glassware?
7.1.5.4				Execution of proper sample preparation?
7.1.5.5				Execution of proper standard preparation?
7.1.5.6				Analysis of proper samples?
7.1.5.7				Execution of proper analytical technique such as mixing, shaking, centrifuging, filtering weighing, and dilution?
7.1.5.8				Use of proper equipment?
7.1.5.9				Use of properly qualified and calibrated equipment?
7.1.5.10				Special considerations for proper execution of HPLC methods?

(*Continued*)

Checklist   (*Continued*)

Step	Yes	No	Na	Laboratory Investigations Subelement (LI)
7.1.5.11				Special considerations for proper execution of GC methods?
7.1.5.12				Special considerations for proper execution of dissolution testing?
7.1.5.13				Special considerations for proper execution of IR analysis?
7.1.5.14				Special considerations for proper execution of UV/Vis spectroscopy analysis?
7.1.5.15				Special considerations for proper execution of titrations including Karl Fischer?
7.1.6				Are OOS investigations handled in a manner consistent with the Barr decision and FDA guidance documents?
7.1.7				Does the SOP prevent release of product by continuous testing, retesting, or resampling (e.g., testing into compliance)?
7.1.8				Does the SOP clearly define situations when reanalysis or retesting are appropriate and provide guidelines on how it will be executed?
7.1.9				Does the SOP address special circumstances such as unidentified chromatographic peaks or shifting retention times of known peaks?

Checklist (*Continued*)

Step	Yes	No	Na	**Laboratory Investigations Subelement (LI)**
7.1.10				Are roles and responsibilities for executing laboratory investigations clearly defined in the SOP?
7.1.11				Are all laboratory personnel trained in the proper execution of laboratory investigations?
7.2				*Laboratory Investigations—Execution*
7.2.1				Is supervisor immediately notified of an OOS, deviation, or atypical event?
7.2.2				Are test material (glassware, test solutions, standard solutions, etc.) retained?
7.2.3				Are equipment and instrumentation left in the same conditions as when the event which caused the investigation to occur?
7.2.4				Are investigations performed in a timely manner?
7.2.5				Are the individuals responsible for generating the OOS, deviation, or atypical event involved in the investigation?
7.2.6				Is a concerted effort made to find a root or assignable cause of all laboratory investigations?
7.2.7				Are investigations closed in a timely manner?

(*Continued*)

Checklist (*Continued*)

Step	Yes	No	Na	Laboratory Investigations Subelement (LI)
7.2.8				If the OOS is discovered after product release, is there a system in place to evaluate impact of product which is in the market?
7.2.9				Is the quality assurance unit integrated into the laboratory investigations process?
7.2.10				If laboratory investigations cannot be completed in their allotted time frame, is there a formal mechanism in place to extend their due date?
7.3				*Laboratory Investigations—Documentation*
7.3.1				Is software (e.g., tracking, trending, and reporting) used to complete laboratory investigations?
7.3.2				Has the software been validated?
7.3.3				Is there a standard report or forms, which are used for documenting laboratory investigations?
7.3.4				Are investigations supported with sufficient data to include chromatograms, spectra, raw data, etc.?
7.3.5				Is there a process for implementing corrective and preventive actions upon closure of laboratory investigations?
7.3.6				Are laboratory investigations tracked and trended?

Checklist (*Continued*)

Step	Yes	No	Na	**Laboratory Investigations Subelement (LI)**
7.3.7				Are the trending reports used during annual product reviews?
7.3.8				If the root cause of the investigation is not determined, is a conscious and rational decision on the impact of continued testing made?
7.3.9				Can final investigation reports be understood from someone outside the organization and stand on their own with limited verbal explanation?
9.0				*Additional Items*

EXAMPLE TEMPLATE FOR AN AUDIT SUMMARY REPORT

INTRODUCTION

At the completion of the audit, the findings need to be reviewed, organized, and presented in a coherent format that can be circulated and reviewed by management as well as other individuals within the organization. The output of this process is the audit summary report (ASR).

The audit summary report is a very valuable and useful document, not only from a what-it-costs standpoint, but also from a compliance and business efficiency standpoint. Specifically, if the audit is comprehensive and executed as delineated in the previous chapters, a significant number of labor-hours will be invested and the direct cost from labor alone can be substantial. However, if performed in a proper fashion, this investment in time and effort has value from a CGMP compliance perspective. Namely, you will have a detailed understanding of your current level of CGMP compliance and be able to show a regulatory agency what you know. From a business standpoint, the ASR lays the basis for the most systematic and efficient means to upgrade your level of compliance. Not to mention that noncompliance in general can be very expensive if it results in significant regulatory action.

The process of organizing and reporting the results is a critical phase because it lays the ground work for developing a future corrective and preventive action plan. The greater the effort expended on determining how the data are to be reported, the more effective and straightforward the creation and implementation of the corrective and preventive action plan.

The final form of the audit summary report is determined by the details and logistics of the audit itself. However, the general structure of all audit summary reports should essentially be the same. The basic components of an audit summary report should include:

Header

The header should identify your facility name and location in addition to all of the personnel involved in the audit. A statement as to the confidential nature of the material included in the report should be made as well.

Background

The background section summarizes the purpose for performing the audit. For example, the audit may be in preparation for an FDA preapproval inspection (PAI), an upgrade of your existing quality systems, or a continuation of an existing audit program.

Approach

This section should describe all the subelements of the laboratory quality management system reviewed during the audit. As described in previous sections these elements include:

- 1.0 Laboratory managerial and administrative systems (MS)
- 2.0 Laboratory documentation practices and standard operating procedures (OP)
- 3.0 Laboratory equipment qualification and calibration (LE)
- 4.0 Laboratory facilities (LF)
- 5.0 Methods validation and technology transfer (MV)
- 6.0 Laboratory computer systems (LC)
- 7.0 Laboratory investigations (LI)

The approach section should also discuss the personnel who were involved in the audit, the mechanics of the audit (e.g., use of checklists), and how the findings were documented (e.g., in a notebook with subsequent documentation on an official finding form, such as a LAF).

Description of Report Format

The body of the report includes sections discussing how the summaries of each of the subelement findings, which are contained within the report, are organized. Namely:

- A brief description of the subelement
- An overview of the current practice at the site or each subelement
- A listing of site documents reviewed
- Gaps in the subelement versus checklists or similar quality review documents
- Additional gaps not correlated to checklists or similar quality review documents
- Potential root causes for the gaps
- Potential corrective action needed to become compliant with CGMPs
- A summary matrix for the above steps, which can be used to creating the corrective action plan

The format of the report can be tailored to fit the individual site needs. However, it is strongly suggested that a summary matrix be included for each subelement. This format greatly enhances the generation of a corrective action plan.

Summary of Results

The summary or results section should capture the total number of findings discovered for all subelements during the audit. As with the individual subelement findings, the summary results should also be organized into a matrix.

Future Work

The future work section should review the steps required for the implementation of a complete audit, namely:

- Preparation phase
- Audit and Data Capture phase
- Reporting phase
- Corrective Action phase

- Verification phase
- Monitoring phase

Some explanation should be given as a need to continue with the Corrective Action phase, and the potential resources, which may be required to complete the full audit.

Laboratory Controls Subelement Sections (Report Body)

This section then presents the data for each subelement as described in the format section, namely:

- Description of subelement
- Current practice
- Site documents reviewed
- Gaps in the system versus audit checklist

The level of detail and breadth of discussion depends upon individual site organizational structure and level of compliance with CGMPs.

Attachments and appendices may be included enhancing the overall readability or usability of the report. Remember, the ASR is used as the basis for corrective and preventive actions and should therefore be as descriptive as possible.

With the previous suggestions in mind, an example report is shown. This report contains all the sections described here and can be modified to suit the needs of the individual organization. As with the checklists however; the example template report shown below is fairly comprehensive but may not be totally inclusive of all the sections required for a specific organization.

YOUR COMPANY QUALITY OPERATIONS LABORATORY

AUDIT SUMMARY REPORT

CONFIDENTIAL: This document is not to be distributed or copied except with the written permission of Your Company Products, LLC Quality Operations

Quality Management System:	Laboratory Control System
Facility Name and Location:	Dummy Products, LLC Quality Operations Laboratory Your Site Operations, Your Site, Your State, 99771
Auditors:	J. Casey, R. Danny, J. Felix, J. Foosball, M. Gummy, A. Lavio, W. Link, R. Metz, A. Quinones, U. Smith, J. Smyth, J. Smooter, E. Vazquez, D. Blistex, V. Dooby, R. Gillen, T. Johnson, B. McMillan, N. Ran

Background

As part of its continuing commitment to quality, Your Company Products, LLC has agreed to voluntary periodic CGMP inspections by the REGULATORY AGENCY. In order to prepare for an inspection, which was originally scheduled for mid-to-late April 2008, Your Company Operations (SITE) Site Management determined the need to conduct a series of self audits. These self audits were to serve two purposes. First, they would be used to prepare for the REGULATORY AGENCY visit. Specifically, a comprehensive review of internal systems would reveal any remaining potential deficiencies with respect to CGMPs and operations in general and allow sufficient time to address any shortcomings. Second, they would provide an excellent opportunity to instruct laboratory personnel with the help of quality assurance on the quality management systems-based audit approach, which was recently formally adopted by the REGULATORY AGENCY. Site management determined that at the beginning of these self audits that the Quality Operations Laboratory would serve as the starting point for this process. Audits of the additional Quality Management Systems (QMSs) would follow as time progressed. It should also be noted that the self-audit approach is meant to lay the groundwork for a formal on-going self-audit–corrective action program which will be spearheaded by the Quality Assurance Department.

Laboratory Operations Self Audit Approach

This quality management system (QMS) self audit reviewed all the current good manufacturing practices (CGMP) systems and practices in the Your Company Operations (SITE) Quality Operations Laboratory. In addition, many of the administrative systems and practices, which can ultimately impact compliance with CGMPs, were also evaluated. The Laboratory Quality Management System encompasses a variety of subelements that cover all aspects of the laboratory. Because the laboratory is staffed as a semi-independent organization, many of the subelements overlap with the other QMSs. Examples include investigations, validation, facilities, training, etc. However, this audit focused on conditions as they exist in the laboratory and when processes connect to organizations outside the laboratory, seeks to assure that the interface is adequate to allow these interdepartmental processes to be conducted seamlessly. The subelements that comprise the Laboratory Operations QMS, which is also referred to as the Laboratory Control System, are:

- 1.0 Laboratory Managerial and Administrative Systems (MS)
- 2.0 Laboratory Documentation Practices and Standard Operating Procedures (OP)
- 3.0 Laboratory Equipment Qualification and Calibration (LE)
- 4.0 Laboratory Facilities (LF)
- 5.0 Methods Validation and Technology Transfer (MV)
- 6.0 Laboratory Computer Systems (LC)
- 7.0 Laboratory Investigations (LI)

The laboratory operations audit team was composed of representatives from SITE QA, the Quality Operations Laboratory, and supervised overall by the senior manager of the Your Company Quality Operations Laboratory. The audit team was divided into 7 subteams, which mirrored each of the subelements listed above. Each subteam was responsible for assessing its specific subelement in the QC laboratories versus the audit checklist for the laboratory. The audit checklist is a comprehensive and detailed document, which is used to systematically evaluate an organization's level of compliance with CGMPs. It represents numerous personnel-years of experience acquired by assisting companies to comply fully with CGMPs.

Deficiencies versus the audit checklist were documented on laboratory audit forms (LAFs). LAFs are considered the raw data captured during the review. LAFs are identified via a standard alpha number naming scheme. For example, **SITE-MS-1.2.2-001** is identified in the following fashion:

- **SITE** = Indicates the deficiency for Your Company Operations (SITE)
- **MS** = Indicates that this is related to the Laboratory Managerial and Administrative Systems (MS) as designated on the audit checklist
- **1.2.2** = Links directly to the audit checklist STEP 1.2.2 which asks "Are training requirements clearly documented in a SOP or similar guidance document including managers, supervisors, analysts and temporary staff?"
- **−001** = Indicates that this is the first finding for this STEP.

Narrative details of the finding are documented on the LAF form itself. Once captured, LAF data are entered into a database, which is used to support corrective and preventive actions (CAPAs), via a corrective action project plan (CAPP). Considerable effort has been made in design of this form to link and consolidate all LAF findings to other quality management systems, and previously documented findings, such as FDA 483 observations, and previously conducted site internal assessments. All LAFs are stored in separate binders corresponding to their subelements. Also included in the binders are a summary of the LAFs for that subelement and the completed audit checklist. The original LAFs (revision 0) have also been scanned or printed to Portable Document File format (*.pdf) and burned as a permanent record to CD-ROM disk.

It should be noted that items, which are not specifically covered by steps in the audit checklist are identified by an "Additional Items (9.0)" designation. For example, SITE-MS-9.1-001 is the first additional deficiency identified by an auditor for issues related to Managerial and Administrative Systems, but not specifically covered by a step number on the checklist. Details of the finding are documented on the LAF as before, and entered into the database. Details of the audit and LAF generation process are show in Appendix A of this report.

This Quality Management System applies to the entire Quality Operations Laboratory including the following sections: Immediate Release and Extended Release Laboratories, Analytical Technical Services, Stability, and Raw Materials. These five sections are located in nine major laboratories in a single building at the Your Company, Your Company Operations, Your Company, Your State, USA. Supervisors report to section managers who in turn report the Quality Operations Laboratory senior manager, who is the laboratory director. Testing responsibilities include: in-process testing, testing to support investigations and stability testing, and raw materials testing for all pharmaceutical solid dosage forms and products which are manufactured, used, or maintained at Your Company at Your Company Operations.

Report Format

The sections shown in the following text are summaries of each of the subelements assessed versus the audit checklist. The format includes: (1) A brief description of the subelement, (2) An overview of the current practice at SITE for each subelement, (3) A listing of site documents reviewed, (4) Gaps in the subelement versus the audit checklist and additional gaps not correlated to the checklist, (5) Potential root causes for the gaps, and (6) Potential corrective action to become compliant. Steps 4–6 are summarized in a matrix.

Summary of Results

The breadth and extent of the quality operations laboratory self audit were extensive. All laboratory personnel who provided information or were interviewed were forthcoming and enthusiastically engaged in the audit process. Moreover, they frequently demonstrated their knowledge of the importance of CGMPs and the need for continuous improvement.

Due to the comprehensive nature of the audit, a good number of gaps were documented. Many of these gaps are not considered *critical*; they would not result in Form 483 observations. However, many of the *noncritical* gaps have to do with administrative systems and practices and can ultimately lead to degraded compliance with CGMPs. Therefore, many of the observed gaps should offer suggestions on "How can this be done better?" It should also be noted that the REGULATORY AGENCY or REGULATORY AGENCY auditors would never have such unfettered access to personnel and records, and would therefore be less likely to document as many findings as was done during this self audit. It should also be noted that several gaps may be related to the same root cause and thus the total number of unique gaps may be less than the number stated.

Table 1 below summarizes the gaps versus the subelements. Table 2 shows the correlation of critical (e.g., = potential 483) gaps and the noncritical gaps (e.g., = can be done better versus the subelements).

TABLE 1

Subelement	# of Checklist Item Gaps	# Non Checklist Gaps	Total # of Gaps	% of Total Gaps Found
1.0 Laboratory Managerial and Administrative Systems (MS)	20	32	52	19.5
2.0 Laboratory Documentation Practices and Standard Operating Procedures (OP)	25	78	103	38.6
3.0 Laboratory Equipment Qualification and Calibration (LE)	25	2	27	10.1
4.0 Laboratory Facilities (LF)	26	0	26	9.7
5.0 Methods Validation and Technology Transfer (MV)	2	35	37	13.9
6.0 Laboratory Computer Systems (LC)	9	0	9	3.4
7.0 Laboratory Investigations (LI)	10	3	13	4.9
Total =	**117**	**150**	**267**	**100**

TABLE 2

Subelement	Total # of Gaps	# of Critical Gaps*	# of Noncritical Gaps
1.0 Laboratory Managerial and Administrative Systems (MS)	52	14	38
2.0 Laboratory Documentation Practices and Standard Operating Procedures (OP)	103	14	89
3.0 Laboratory Equipment Qualification and Calibration (LE)	27	10	17
4.0 Laboratory Facilities (LF)	26	11	15
5.0 Methods Validation and Technology Transfer (MV)	37	3	34
6.0 Laboratory Computer Systems (LC)	9	4	5
7.0 Laboratory Investigations (LI)	3	0	3
Total =	**267**	**59**	**208**

*Critical = could potentially warrant a Form 483 observation from FDA.

Future Work

The completion of this report represents the completion of the first three steps in a complete self-audit process, which includes the following phases:

- Preparation phase
- Audit and Data Capture phase
- Reporting phase
- Corrective Action phase
- Verification phase
- Monitoring phase

To complete the process, this report should be used to create a comprehensive corrective action project plan (CAPP) which will be used to implement corrective and preventive actions (CAPAs). This should in turn be followed by implementation of a verification plan, which will be integrated to a monitoring plan that includes periodic reassessments and reporting of those results. Appendix B outlines the process from LAF generation to CAPP implementation in detail.

Laboratory Control System Subelement:	*1.0 Laboratory Managerial and Administrative Systems (MS)*
Auditor(s):	J. Felix, J. Smooter, D. Blistex

Description of the QMS Subelement 1.0 Laboratory Managerial and Administrative Systems (MS)

The Laboratory Managerial and Administrative Systems subelement has eight individual topics as defined in the audit checklist. These are: Organizational Structure and Roles and Responsibilities, Training, Tracking and Trending—Statistical Quality Control, Complaints, Laboratory Purchasing and Requisition, Laboratory Administration, Laboratory Chemicals, Solutions, Reagents, and Supplies, and Laboratory Reference Standards and Solutions. Each of these eight topics is addressed separately as part of the subelement discussion in the following sections.

Current Practice 1.1 Organizational Structure and Roles and Responsibilities

This review involved conducting interviews with supervisors, managers, and personnel within the Quality Operations Laboratory. In many cases the interviews are the result of a "follow the sample" approach to auditing. That is, personnel were asked to track a sample from receipt to final disposition.

The roles and responsibilities for each position in the Quality Operations Laboratory are defined by a combination of organizational charts, standard operating procedures, resumes, position descriptions, training qualifications, and yearly reviews. Section supervisors assign work responsibilities based on their understanding of the workload and knowledge, experience and abilities of the scientists to perform the analysis. Signature authority and responsibility are not clearly defined by SOPs. However, the supervisors and managers have sufficient knowledge of their tasks and authority to identify the occurrence of departures from the SOP. For example the supervisors can perform investigative testing of suspect samples to determine an assignable laboratory cause. The staff has turned over significantly over the last 2–3 years, and considerable effort has been made to bring the laboratory to a higher state of compliance with CGMPs. This has increased the workload significantly and results in a substantial number of the supervisors and managers working 10–15 hours of overtime on average per week.

Current Practice 1.2 Training

This review involved discussions with a newly assigned laboratory training manager. Since the Quality Operations Laboratory operates as a semi-independent entity, the training manager is responsible for coordinating, documenting, and in many cases preparing/conducting the majority of laboratory training. Technique and product specific training is conducted directly by this individual or by subject matter experts within the lab. Training is documented in several ways, including course attendance records, completed knowledge checks, and data and instrument outputs for hands-on procedure specific training. These documents are tracked via manual systems like the internally generated training matrices for hands-on training and by the XTrain software package. These tools are used to create a training file (binder) for each person in the laboratory. These binders are stored at various locations within the lab, usually close to the primary work location of the individual. No one system compiles and tracks training documentation, which is currently a work topic for the training manager. This is part of the overall effort by the training manager to form a more coherent and effective training program for laboratory personnel, which will include generating a master training, schedule, formally codifying all training modules, and comprehensive use of XTrain to track all laboratory training.

Current Practice 1.3 Tracking and Trending-Statistical Quality Control

These duties are not the responsibility of the Quality Operations Laboratory and are performed by the Quality Management Group, which falls under the auspices of the Quality Assurance Unit. Therefore, an audit of this system was not performed at this time, but will be included in future, expanded audits of the additional Quality Management System (QMS).

Current Practice 1.4 Complaints

Complaints are received at the Tahiti Regulatory Department and forwarded to the SITE Regulatory Department as necessary. The Quality Operations Laboratory is responsible for conducting only the required testing determined by SITE Regulatory. Testing is requested and initiated by the SITE Regulatory Department via issuance of Attachment II (a form) found in YLP 05-011 *Complaint Investigation Report*. The laboratory uses this form and follows YLP 02-055 *Investigating Customer Complaint Samples* to generate the results and forward it back the Regulatory Department. The lab is not involved in any decision making process, only data generation and reporting as delineated in these procedures.

Current Practice 1.5 Laboratory Purchasing and Requisition

This review involved discussion with the analytical services supervisor who has recently been placed in charge of laboratory purchasing and requisitions. Previous to this, it had primarily been the responsibility of the stock room supervisor and instrumentation supervisor, with final signature approval performed by the senior manager of the laboratory. To obtain supplies, laboratory personnel request needed laboratory supplies from the laboratory stockroom supervisor. The stockroom supervisor then completes a purchase order, obtains the analytical services manager's or the senior manager's authorization and then forwards the requisition to the purchasing function. Levels of signature authority are not clearly defined in writing. There is no indication if the proper grade of reagents is taken into account during the requisition process. Although QA has a list of qualified vendors, the laboratory does not and may not take this into consideration when making requisitions. Monitoring of expenditures and budget generation and review has primarily been the responsibility of the senior manager but is shifting to the analytical services supervisor. None of these individuals has received significant formal training on theses topics.

Current Practice 1.6 Laboratory Administration

This review involved interviews with the senior laboratory manager, analytical services manager, training manager, and the instrumentation supervisor using a prepared questionnaire and the audit checklist. To this point most of primary administrative tasks for the laboratory were spread out among the managers, and instrumentation supervisor. Due to work schedules and volume many of the issues addressed in the audit checklist were never addressed in a coordinated, definitive fashion. Because of this the senior manager has recently suggested the creation of a formal position within the laboratory where a single individual will be responsible for most of the diverse administrative issues such as budget management, overseeing stockroom operations, capital expenditures, managing external contracts, etc.

Current Practice 1.7 Laboratory Chemicals, Solutions, Reagents and Supplies

This review involved touring the laboratories, reviewing existing SOPs and answering questions on the audit checklist related to laboratory chemicals, solutions, reagents and supplies. SOPs YLP 02-013 *Maintaining Volumetric and Reference Standard* Solutions and YLP 02-014 *Storage of Reactive Solutions in the Laboratory* give fairly comprehensive instructions with respect to the handling of chemicals, reagents and solutions. When coupled with the USP, EP, and product specific procedures (PSD) they provide sufficient guidance to comply with current industry standards. A spot check of labeling of solutions and reagent in the laboratory confirmed this to be generally true.

Current Practice 1.8 Laboratory References Standards and Solutions

This review involved using the audit checklist and discussions with personnel responsible for receipt, labeling, handling, and recertification of reference standards and materials in the Quality Operations Laboratory. The Central Reference Standard Group in Tahiti has the primary responsibility for reference standards at Your Company. Details of these responsibilities are addressed in the LEVEL II 22,414 *Reference Standards*. Currently there is no Level III SOP addressing reference standards at SITE and the LEVEL II has not been implemented at SITE. Consequently there are shortcomings in the reference standard program at SITE. In house reference standards are received from Tahiti, and compendial standards are ordered and received directly from the source. Once received, all standards are logged and secured with lock and key; however, they are not stored in environmentally controlled environments. In addition, certificates of analysis are not handled as controlled documents. In general, reference standards are not handled in accordance with current industry standards. This area represents one of the greater challenges in the Quality Operations Laboratory.

Site Documents Reviewed

- Blank copy of Your Company professional/managerial performance appraisal
- Chromatography module
- Dissolution test module

- Human Resources organizational charts
- Individual training records (Various Binders)
- Interoffice memorandum *Analyst Qualifications and the Site Training Function Responsibilities, 26 Nov 2007*
- YLP-02-108
- Quality Operations Laboratory Your Company operation products
- Quality Operations Laboratory basic training module
- Quality Operations organizational charts
- Spectrophotometery module
- Training attendance records (Various)
- Training matrices, all sections

Gaps in the System versus the Audit Checklist

The following matrix correlates potential gaps uncovered during the self audit and links them to specific line times in the audit checklist. In addition to the gap, the matrix also indicates if the gap represents part of the system that is in sustainable compliance (e.g., No = a Critical finding which potentially could result in a Form 483 finding if discovered by the REGULATORY AGENCY), what the potential root cause may be for the gap, and some suggestions for potential corrective action to make the system become compliant. If the auditor did not make suggestions as to the root cause or potential corrective action to become compliant, the statement "None offered" is included in the space.

In addition to the findings correlated directly to the audit checklist, additional gaps are included in the matrix. In some circumstances the description serves as the gap and therefore the gap block may state "Same as description."

Gaps in the System: Laboratory Subelement 1.0 Laboratory Managerial and Administrative Systems (MS)

Checklist Item Number and Description	Gap	In Substantial Compliance?	Potential Root Cause	Potential Corrective Action to Become Compliant
1.1.6 What is the ratio of supervision to analyst? (8:1 recommended)	The agreed target ratio of analyst to supervisor is 8:1. That ratio has not yet been achieved in the stability area.		Some laboratory personnel have recently left the company due to a major hiring initiative from another local firm.	Evaluate the root cause of departure. Reevaluate the current compensation system. Make sure to include noncash incentives as part of the evaluation.
1.1.9 Are signature authority, responsibility, and accountabilities appropriately and clearly defined?	The Quality Operations Laboratory does not have a memorandum or similar document, which delineates signature authority for laboratory managers, supervisors, or other personnel.	No	Many of the administrative tasks assigned to managers and supervisors are not part of their formal job descriptions and done on-the-side in conjunction with their regular duties.	Create a document which clearly defines signature authority. Create a position within the laboratory where the individual is responsible for the general administrative functions such as training, supply requisition (e.g., stockroom), equipment maintenance and calibration, etc. Then assign this individual the responsibility for creating, submitting, reviewing, and monitoring the training budget by interacting with the managers and supervisors. This person

(Continued)

159

Gaps in the System: Laboratory Subelement 1.0 Laboratory Managerial and Administrative Systems (MS) *(Continued)*

Checklist Item Number and Description	Gap	In Substantial Compliance?	Potential Root Cause	Potential Corrective Action to Become Compliant
				will have most of the signature responsibilities in the laboratory.
1.1.10 Do systems exist to enhance communications, understanding, and working relationships between the laboratory and quality assurance personnel?	Systems do not currently exist to enhance communications, understanding, and working relationships between the laboratory and QA personnel. In general, QA is not involved in the over site of day-to-day operations of the laboratory.	No	None offered.	Laboratory management needs to perform an analysis of where QA personnel can add the most value within the Quality Operations Laboratory organization. Once the analysis is complete, discussions need to be initiated with QA to determine the best path forward.
1.1.11 Does a personnel performance system exist which tracks laboratory personnel strengths and weaknesses and establishes corrective action procedures to mitigate weaknesses?	The Personnel Performance Evaluation System as currently used by the laboratory may not be effective in identifying employee strengths and weaknesses, identifying avenues for continuous improvement, and providing continuous and timely feedback on performance.		Existing performance evaluation system is not metrics based.	To augment the existing system, initiate a complimentary performance metrics based system to evaluate performance in the laboratory.

Item	Description	Status	Finding	Recommendation
1.1.12 Does a master testing schedule or similar document(s) exist to insure smooth workflow, and minimize laboratory personnel over commitment?	The laboratory does not currently have a master testing schedule to insure smooth workflow and minimize laboratory personnel over commitment.		Communications between the laboratory, manufacturing, and technical services needs to be improved.	Design and development of proper communication systems between these groups and supervise their implementation.
1.2.2 Are training requirements clearly documented in an SOP (including managers, supervisors, analysts, and temporary Staff)?	There is currently no SOP in place to address revisions of training curricula, or clearly define training requirements for all personnel including managers, supervisors, analysts, and temporary staff.	No	Training supervisor is a recently created position. In addition, a specific written path forward to develop department training objectives has not been created.	As part of the CAPP, a performance matrix needs to be created and executed for the training supervisor. This matrix should include the creation of an SOP to address the issues highlighted here.
9.1 The laboratory training supervisor is not involved in the interview and hiring process of laboratory personnel.	Same as description		There is currently no formalized system in place which addresses all of the components required to properly search for, interview, hire, train, and evaluate personnel.	The process for hiring new personnel within the Quality Operations Laboratory needs to be reevaluated to include not only steps for assessing the needs of the department, but also what the impact of hiring the individual has on the workflow and resources needed to add this individual to the

(Continued)

Gaps in the System: Laboratory Subelement 1.0 Laboratory Managerial and Administrative Systems (MS) (*Continued*)

Checklist Item Number and Description	Gap	In Substantial Compliance?	Potential Root Cause	Potential Corrective Action to Become Compliant
				department roles. An equivalent system of DQ/IQ/OQ/PQ and maintenance and calibration, which is used in purchasing, qualifying, and installing equipment needs to be developed and implemented for personnel.
9.2 Initial SOP training and training on revisions to SOPs has limited effectiveness.	Same as description		Current system for SOP training and SOP revisions training is based on an augmented "read-understand-test" scenario. Studies have shown this is perhaps the least effective means of training.	Design and develop a multifaceted approach to SOP training, which includes classroom, video, self-instruction, and bench-chemist led instruction.
9.3 The laboratory training Supervisor does not have full access to the XTrain computer software system.	Same as description		Training responsibilities for the QC Laboratory are currently split between the Regulatory Department and	Arrange with the Regulatory Department for the QC training supervisor and general administrative

		the laboratory training supervisor. However, accountability, authority, and responsibility are not clearly delineated leading to potential shortcomings, such as access to XTrain.	manager to assume total accountability, authority, and responsibility for the laboratory training program to include access to the XTrain system.
9.4 Training for raw material methods and USP methods are not currently tracked by the XTrain system or the augmenting paper system. They are contained in training grids, which is a third way that training is tracked.	Same as description		Same as 9.3
9.5 Records in the XTrain system are difficult to access in a timely manner.	Same as description	XTrain resides on a corporate computer server located in the Mainland USA. Therefore, large-scale data transfer, which is required to query and review files, takes a significant amount of time.	Work with the owners of the corporate computer server and IT to determine if speed can be improved. If it is not possible, determine whether a copy of XTrain can reside locally at SITE.

Laboratory Control System Subelement:	2.0 Laboratory Documentation Practices and Standard Operating Procedures (OP)
Auditor(s):	J. Foosball, M. Gummy, A. Quinones, D. Blistex

Description of the QMS Subelement 2.0 Laboratory Documentation Practices and Standard Operating Procedures (OP)

The Laboratory Documentation Practices and Standard Operating Procedures subelement has nine individual topics as defined in the audit checklist. These are: SOPs–General, SOPs–Specific Procedures, Laboratory Test Procedures, Laboratory Data and Results, Security of Data, Distribution of Results, Chromatography, In-Process Testing, and Assignment of Retest/Expiry Dates. Each of these nine topics is addressed separately as part of the subelement discussion in the following sections.

The review for the entire Laboratory Documentation Practices and Standard Operating Procedures subelement involved an in-depth interview with the laboratory documentation supervisor. In addition to using the audit checklist the supervisor was asked to explain, in detail, the major and minor facets of his/her job and to process diagram workflow in circumstances where it was appropriate.

Current Practice 2.1 SOPs—General

This section review focused on ascertaining whether the laboratory has: (1) The proper documents on hand to complete the tasks and provide guidance, (2) These documents are clearly written and used appropriately by laboratory personnel, and (3) The proper systems for creating, revising, and storing these documents are in place. In general it was found that the status of document control was sufficient but that the quality of documents was in some cases lacking. It was discovered that many procedures were difficult to follow as written and in some cases had errors which could adversely affect results. Although the documents are controlled appropriately, much of this system is manual and very labor intensive.

Current Practice 2.2 SOPs—Specific Procedures

This section review focused on ascertaining whether specific procedures related for common laboratory operations were in place. In most cases procedures were in place; however, SOPs on computer validation, glassware cleaning, and document review were not.

Current Practice 2.3 Laboratory Test Procedures

This section review focused on reviewing specification documents, product testing procedures, and instrument use procedures and control of these documents. As with the general observations, the instrument procedures often lacked sufficient detail making them difficult to use as written. Also, the creation and revision process for these documents is via manual system and is cumbersome and inefficient.

Current Practice 2.4 Laboratory Data and Results

This section review focused on ascertaining whether laboratory data and results are properly captured, processed, reviewed, and stored for later retrieval. In general, the laboratory has good systems for addressing these issues. Data are captured via instrument outputs and/or bound notebook entry. Calculations are preformed by software algorithms or by manual means. Regardless, all data are reviewed and signed off by a data verifier. Some issues do exist with respect to standardized integration/reintegration procedures for HPLC and GC chromatograms.

Current Practice 2.5 Security of Data

This section overlaps with the section described above for data and results in general. With the diversity of the tasks performed in the laboratory (e.g., finished product testing on one end and methods validation on the other) there are different means and locations for the storage of data. Finished product data are stored in the SITE's central archive. Although very secure, access to these data is very restricted making it difficult for someone to easily retrieve them. Methods validation data are at the other extreme and maintained in the laboratory area in standard locked filing cabinets. Open access, to just about anyone in the laboratory, is available during the work day and strict check in and out procedures are not followed. These records are also susceptible to water (from the overhead sprinkler system) and fire damage. Overall there is no disaster recovery system in place and no offsite data and records storage facility exists.

Current Practice 2.6 Distribution of Results

Much of the audit checklist review for this section focuses on the use of LIMS in the laboratory. SITE Quality Operations is not currently using a LIMS system so most of these questions are not applicable. Transcribed data that are captured via instrument output and transcribed to paper are properly reviewed by a second party.

Current Practice 2.7 Chromatography

This section addresses the proper use of system suitability for HPLC and GC chromatographic runs. The SITE Quality Operations Laboratory establishes appropriate system suitability for each of their chromatographic runs.

Current Practice 2.8 In-Processing Testing

All in-process testing is performed by the In-Process Quality Testing Laboratory.

Current Practice 2.9 Assignment of and Retest/Expiry Dates

All issues related to retesting and expiry is addressed by the Product Disposition Department.

Site Documents Reviewed

- C-141 Mesh Analysis, Potassium Chloride
- CV-064
- F-137 P-Dip Friability
- K-191
- YLP-02-002
- YLP-02-004
- L128 Theosux Loss on Drying
- MV-02-111
- MV-02-112
- Notebook YLP-2470 pp. 38
- Notebook YLP-2474
- Notebook YLP-2484 pp. 50
- Notebook YLP-2506 pp. 86
- Notebook YLP-2510 pp. 4-11
- Notebook YLP-2512 pp. 80-83
- Notebook YLP-2513 pp.139-147, 140-147
- P-226 P-Dip Chloride Identification
- PSD ??? V3 Chicken Soup Determination of Degradation Products
- PSD 2852 V8 Veggie Soup Content Uniformity
- PSD 2852 V8 Veggie Soup Description
- PSD 2885 V8 Veggie Soup Dissolution
- PSD 30503 Identification, TLC Micronized Loratadine
- PSD 4703 V3 Chicken Soup Description
- PSD 4703 V3 V3 Chicken Soup Description
- STP-591 Rowboat Moisture Content
- STP-688 Moisture in Vanilla
- USP <461> Nitrogen Content (in Cross Povidone)
- USP <578> P-Dip Disintegration
- USP <905> Rowboat Content Uniformity
- LEVEL II 22,409

Gaps in the System versus the Audit Checklist

The following matrix correlates potential gaps uncovered during the self audit and links them to specific line times in the audit checklist. In addition to the gap, the matrix also indicates if the gap represents part of the system that is in sustainable compliance (e.g., No = a Critical finding which potentially could result in a Form 483 finding, if discovered by the REGULATORY AGENCY), what the potential root cause may be for the gap and some suggestions for potential corrective action to make the system become compliant. If the auditors did not make suggestions as to the root cause or potential corrective action to become compliant, the statement "None offered" in included in the space.

In addition to the findings correlated directly to the audit checklist, additional gaps are included in the matrix. In some circumstances the description serves as the gap and therefore the gap block may state "Same as description."

Gaps in the System: Laboratory Subelement 2.0 Laboratory Documentation Practices and Standard Operating Procedures (OP)

Checklist Item Number and Description	Gap	In Substantial Compliance?	Potential Root Cause	Potential Corrective Action
2.1.3 Are the SOPs current, clearly written, and accessible to all appropriate personnel?	Many SOPs and standard test procedures are not clearly written and therefore difficult to understand and follow.		None offered.	Perform a detailed, logical, prioritized wholesale review of existing SOPs within the laboratory. Consider clarity and language consistency in the review.
2.1.4 Is there a system for periodic review of all SOPs to assure that they are consistent with current company and industry practices?	Although there is a system in place, it is manual and very difficult to manage efficiently.		There are too many SOPs in use to manually track and review.	Begin using Wise Crack or similar software to organize and track periodic review of SOPs.
2.1.6 Is there a system for controlling the issuance and revision of all SOPs?	The system for controlling the issuance and revision of all SOPs is not fully documented in an SOP.	No	None offered.	Revise existing SOPs covering change control in the laboratory and include all steps which are currently performed but not documented.
2.1.7 Are policies and manuals used that supplement the SOPs? (e.g., Level I and II guidance documents)?	Policies and manuals do exist such as the ASQ and SQA documents, but they are not accessible or current.		None offered.	Revise existing SOPs covering change control in the laboratory and include all steps which are

Question				
				currently performed but not documented.
2.1.9 Are all SOPs reviewed and updated at least every year?	No	There is currently no automated system to track the mandatory 1-year SOP review requirement.	None offered. SOPs should be reviewed at least once a year.	Revise existing SOPs covering change control in the laboratory and include all steps, which are currently performed but not documented. Formalize and organize, using software tools when possible, the 1-year review period.
2.2.1 Are there specific standard operating procedures covering: change control?	No	There is currently no laboratory procedure which, address the change authorization process.	None offered.	Create a document which delineates the change authorization process for the laboratory.
2.2.8 Are there specific standard operating procedures covering: Policy for identification and reporting new impurities and/or higher levels of previously known impurities?	No	There are no SOPs addressing discovery of new impurities and/or higher levels of known impurities such as what is seen in Chicken Soup.	None offered.	Create a document which delineates how to address new impurities and/or higher levels of known impurities when they appear during stability studies.

(Continued)

Gaps in the System: Laboratory Subelement 2.0 Laboratory Documentation Practices and Standard Operating Procedures (OP) (*Continued*)

Checklist Item Number and Description	Gap	In Substantial Compliance?	Potential Root Cause	Potential Corrective Action
9.17 Data recording practices are different for the Analytical and In-Process laboratories (controlled data sheets vs. notebooks). Practices should be harmonized.	Same as description		The In-Process Laboratories often function independently from the other labs within the QO laboratory operation. This may lead to differences in practices between the Chemistry and the In-Process labs.	None offered.
9.18 No system exists to generate Standard Test Procedures within a reasonable period of time following completion of methods validation activities. For example, the Stickmud Traces method has been validated since (MV 69-007) 1999, but there is still no written procedure.	Same as description		Current change authorization procedure is inefficient and needs to be revised.	Perform a detailed, process diagram-based evaluation of the CA procedures. Modify to improve efficiency.

Laboratory Control System   *3.0 Laboratory Equipment*
Subelement: *Qualification and Calibration (LE)*

Auditor(s):   J. Felix, A. Lavio, J. Smyth, N. Ran

Description of the QMS Subelement 3.0 Laboratory Equipment Qualification and Calibration (LE)

The Laboratory Equipment Qualification and Calibration subelement has three individual topics as defined in the audit checklist. These are: Laboratory Equipment Procedures—General, Laboratory Equipment Procedures—Specific, and Laboratory Equipment Procedures—Computer Controlled. Each of these topics are addressed separately as part of the subelement discussion in following sections.

Current Practice 3.1 Laboratory Equipment Procedures—General

This section addressed basic aspects of the laboratory equipment qualification, calibration, and maintenance program. Some of the review items included verification of master equipment lists, procedures for maintenance and calibration, equipment-use logs, and labeling of equipment. Overall the Quality Operations Laboratory has the rudimentary components of this system in place although the personnel are overburdened with other aspects of their jobs. This leads to a degraded state of compliance with their own procedures and industry practice in general.

Current Practice 3.2 Laboratory Equipment Procedures—Specific

This section reviewed aspects of equipment IQ, OQ, PQ, calibration and maintenance for specific pieces of equipment one would find in a typical analytical laboratory. Equipment records reviewed included pH meters, balances, thermometers, UV spectrophotometers, dissolutions baths, HPLCs, and GCs. As stated here, the Quality Operations Laboratory has the rudimentary components of this system in place although the personnel are overburdened with other aspects of their jobs. This leads to a degraded state of compliance with their own procedures and industry practice in general.

Current Practice 3.3 Laboratory Equipment Procedures—Computer Controlled

This section reviewed specific aspects of equipment IQ, OQ and PQ, related to computer-controlled equipment and the appropriateness of software validation. As with the reference standards handling, in general, equipment qualification of computer controlled devices is not handled in accordance with current industry standards. This area also represents one of the greater challenges in the Quality Operations Laboratory.

Site Documents Reviewed

- YLP 11-028 *Lab Hood Verification*
- YLP 11-*122 Operation, Qualification and Calibration of Balances*
- YLP 011-149 *Operation of the Autosampler System For Total Organic Carbon (TOC) Analyzer Sievers, Model 800*
- YLP 011-174 *Operation of the Amsco SV-120*
- YLP 02-002 *Laboratory Safety Procedures*
- YLP 02-007 *Analytical Laboratory Investigations Including Out-of-Specification Results*
- YLP 02-016 *Requalification of Reference Standards*
- YLP 02-105 *Verification of Calibration Certificates Issued by an Outside Contractor*

Gaps in the System versus the Audit Checklist

The following matrix correlates potential gaps uncovered during the Self Audit and links them to specific line times in the Audit Checklist. In addition to the gap, the matrix also indicates if the gap represents part of the system that is in sustainable compliance (e.g., No = a Critical finding which potentially could result in a Form 483 finding, if discovered by FDA), what the potential root cause may be for the gap, and some suggestions for potential corrective action to make the system become compliant. If the auditors did not make suggestions as to the root cause or potential corrective action to become compliant, the statement "None offered" is included in the space.

In addition to the findings correlated directly to the audit checklist, additional gaps are included in the matrix. In some circumstances the description serves as the gap and therefore the gap block may state "Same as description."

Gaps in the System: Laboratory Subelement 3.0 Laboratory Equipment Qualification and Calibration (LE)

Checklist Item Number and Description	Gap	In Substantial Compliance?	Potential Root Cause	Potential Corrective Action
3.1.1 Are the laboratories equipped with all of the necessary instruments for the analytical testing to be performed?	Many instruments are awaiting IQ/OQ/PQ, and are therefore not in use. Work around situations are present in the lab area.		There is a lack of adequate laboratory resources for performing IQ/OQ/PQ. There may be a lack of adequate GMP training on equipment related issues.	Conduct an inventory of equipment available in the labs and compare this inventory with the needs of the lab. Also, QA should be more actively involved in auditing the labs and associated areas to verify that (1) CGMPs are practiced and (2) Equipment qualifications are completed in a timely manner.
3.1.3 Is there a master equipment list available and is it properly maintained?	Test equipment and lab facilities (hoods) are not included in any equipment list. More than one (1) master equipment list exists. 1. There should be only one version of the master equipment list. 2. Master equipment list should include all analytical instrumentation, laboratory	No	Calibration/metrology program description SOP does not describe how to document and track analytical lab equipment and associated test standards.	Perform an equipment inventory in the labs and in the metrology area. Work with metrology to update and correct the master equipment list. Eliminate all other existing lists. Update associated SOPs with instruction for a single master equipment list.

(Continued)

Gaps in the System: Laboratory Subelement 3.0 Laboratory Equipment Qualification and Calibration (LE) *(Continued)*

Checklist Item Number and Description	Gap	In Substantial Compliance?	Potential Root Cause	Potential Corrective Action
	facilities equipment, and all test equipment and standard reference materials (SRMs).			
	3. Master equipment list must also indicate status (active/inactive equipment).			
3.1.6 Are calibration/service vendors qualified and are their training records available?	Contractors/vendors are not consistently trained to pertinent company SOPs. For example:	No	No QA audit of the vendors has been conducted.	Obtain contractor training records and/or train contractors in SP SOPs; update contractor files. Implement an SOP if necessary.
	1. There is no indication that the site SOP is followed for calibration of AA by the Pooky-Elmer service personnel or if they have been trained on the SOP			
	2. There are no training records for Pooky-Elmer personnel.			
3.1.9 Does each piece of equipment have a logbook or file, which documents instrument	Logbooks were available for the laboratory equipment, but entries were inconsistent. This problem prevailed throughout.	No	Ineffective logbook entry and usage training.	Review logbook entry practices throughout the laboratory. Retrain personnel as necessary.

maintenance, calibration and repair histories?

1. AAA–4 contained no distinct entry indicating the performance of PM.

2. IR-8 had no log-in or log-out indicated in logbook.

3. HPLC-25 was out of service, but the associated logbook had no entry to indicate the out-of-service status

4. Differing formats for time entries have been entered into logbooks.

5. Logbook XX-8 had a yellow sticky note with writing on it adhered to one of its pages. All information should be recorded directly onto the logbook page; other sources for notes are not to be used. This logbook also contained out of sequence dating.

6. The logbook for Dissolution Bath 1 was missing the entry for the calibration of the thermistor.

7. For the Star Ion Analyzer, calibration prior to use is required. The calibration has

(Continued)

Gaps in the System: Laboratory Subelement 3.0 Laboratory Equipment Qualification and Calibration (LE) *(Continued)*

Checklist Item Number and Description	Gap	In Substantial Compliance?	Potential Root Cause	Potential Corrective Action
	not been noted in the associated logbook. 8. HPLC 52 had no entry for calibration. 9. TCB-1 showed no record of use logbook available.			
3.1.12 Has IQ/OQ/PQ been performed and properly documented for all equipment?	Instruments are not in use because IQ/OQ/PQ has not yet been completed. For example, no balance printers have been qualified.		Lack of effective IQ/OQ/PQ strategy.	Mandate a timeline for completion of IQ/OQ/PQ and salvage of lab equipment. Conduct internal audits to verify adherence to this timeline.
3.1.17 Is there an SOP, which requires that instrument failing calibration be removed from service?	YLP 666 should be used for out-of-service labeling. Also, some equipment labelled as out-of-service was still being used. For example: 1. The out-of-service label affixed to HPLC-32 was not the correct form. A piece of paper with "out-of-service" written on it was used to label the HPLC system as out-of-service.		Ineffective or inadequate SOP and GMP training for this subject.	Retrain personnel on associated SOPs and provide refresher GMP training in this area. Confirm that SOPs are adequate. If needed, revise and update SOPs.

	2. EEE-13 is labeled as out-of-service, but it is being used.			
9.1 Issues exist in the glassware washing area.	Issues in the glassware washing area include: 1. There was a big puddle of water on the floor by the water system (Water System #1). 2. The water system is inadequately labeled (masking tape was used to indicate valves). 3. No ID on glassware dryers. 4. The glassware dryers are used for drying critical glassware, however, they are not on a calibration schedule.	No	Lack of an internal audit system.	QA should audit the labs and associated areas to verify: 1. SOPs are adhered to. 2. CGMPs are practiced. 3. Equipment qualifications are completed in a timely manner.
9.2 Issues exist with respect to standard reference materials (SRMs)	Issues include lack of organization of SRMs such as: 1. Test equipment and standard reference Materials are not listed on the master equipment list.	No	Overburdened personnel, and infective GMP training on this subject.	Conduct remedial GMP training and follow-up audits. Hire personnel to keep the metrology lab area organized.

(Continued)

Gaps in the System: Laboratory Subelement 3.0 Laboratory Equipment Qualification and Calibration (LE) (*Continued*)

Checklist Item Number and Description	Gap	In Substantial Compliance?	Potential Root Cause	Potential Corrective Action
	2. DC-06 (dial thickness gauge) had an incorrect calibration sticker on it. The date was mistakenly written so that the piece appeared to be out-of-calibration. There had been no correction yet made and/or it had not been taken out of service. Personnel were planning on using it if they needed it. GMP retraining needs to be done in reference to this issue.			

Laboratory Control System Subelement:	*4.0 Laboratory Facilities (LF)*
Auditor(s):	J. Felix, A. Lavio, U. Smith, T. Johnson

Description of the QMS Subelement 4.0 Laboratory Facilities (LF)

The Laboratory Facilities subelement has three individual topics as defined in the audit checklist. These are: Laboratory Facilities—General, Safety and Environmental Concerns, and Laboratory Glassware. Each of these topics is addressed separately as part of the subelement discussion in the following sections.

The review of this subelement for all of the following sections involved making copies of the audit checklist and systematically inspecting each one of the separate labs and spaces in the Quality Operations Laboratory area. A final *complete* checklist was then filled out and the individual completed checklists were attached for reference.

Current Practice 4.1 Laboratory Facilities—General

For this section the overall physical layout, outfitting, and construction was evaluated for the Quality Operations Laboratory. This included determining the adequacy of space, the adequacy of utilities and services as well as the availability of SOPs and status of general housekeeping. In several cases, the laboratory was lacking in the areas of HVAC and water systems qualification and maintenance.

Current Practice 4.2 Safety and Environmental Concerns

For this section, the overall status of the safety systems in the Quality Operations Laboratory was reviewed. This included evaluation of the safety and environmental SOPs, the status of hood testing, and hazardous waste handling and disposal. Much of this system could be upgraded to meet current industry standards.

Current Practice 4.3 Laboratory Glassware

This review detailed current practices for manual and mechanical glassware washing the Quality Operations Laboratory. At the time of the self audit, the laboratory was in the process of assisting in development of the new LEVEL II 22,777 *Laboratory Volumetric Glassware Requirements and Glassware and Laboratory Equipment Cleaning.* Therefore, no current Level III document exists and some issues with respect to validation of manual and mechanical glassware cleaning exist.

Site Documents Reviewed

- YLP 011-028
- YLP 01-119
- YLP 02-003
- YLP 02-020

- YLP 02-023
- YLP 02-033
- YLP-09-008 rev8
- LEVEL II 10,101

Gaps in the System versus the Audit Checklist

The following matrix correlates potential gaps uncovered during the self audit and links them to specific line times in the audit checklist. In addition to the gap, the matrix also indicates if the gap represents part of the system, that is in sustainable compliance (e.g. No = a Critical finding which potentially could result in a Form 483 finding, if discovered by FDA), what the potential root cause may be for the gap, and some suggestions for potential corrective action to make the system become compliant. If the auditors did not make suggestions as to the root cause or potential corrective action to become compliant, the statement "None offered" is included in the space.

In addition to the findings correlated directly to the audit checklist, additional gaps are included in the matrix. In some circumstances the description serves as the gap and therefore the gap block may state "Same as description."

Gaps in the System: Laboratory Subelement 4.0 Laboratory Facilities (LF)

Checklist Item Number and Description	Gap	In Substantial Compliance?	Potential Root Cause	Potential Corrective Action
4.1.1 Are the physical construction of the laboratory areas adequate for testing and all routine activities with respect to:	The ceiling and walls of room 222 are not the same quality of the manufacturing areas they serve. Also the floor paint is peeling causing housekeeping concerns.	No	Area relocated and not upgraded. Physical facilities are not adequate for testing and other routine activities (room too small).	Replace or upgrade ceiling and walls to meet requirements of the draft LEVEL II. Clean and paint floor. Explore possible relocation.
4.1.1.1 Size 4.1.1.2 Layout and design 4.1.1.3 Sample receipt 4.1.1.4 Appropriate tables for balances/instruments, etc. 4.1.1.7 Data entry, recording, writing areas	In room 327 access to electrical panel is blocked by a storage cabinet. Overall, room 327 is very full thus restricting access to equipment.			Generate a workorder to move balances (scales) from current location to an adequate one. Make purchase order for a marble balance table. Follow SOP YLP 011-122.
4.1.1.8 Sample storage and retention 4.1.1.9 Refrigeration	In extended release lab room 319, a balance is located on a table top.		SOP YLP 011-122 was not followed.	
	In stability room 319, a balance is located under an air supply register.			
	Electrical device panels and receptacles are inconsistently identified in all QO laboratory and manufacturing areas.		No system in place in the engineering area for the identification of electrical devices.	Generate a control system (SOP). See YLP 09-008 a related SOP for major electrical equipment.

(Continued)

Gaps in the System: Laboratory Subelement 4.0 Laboratory Facilities (LF) *(Continued)*

Checklist Item Number and Description	Gap	In Substantial Compliance?	Potential Root Cause	Potential Corrective Action
4.1.2 Are proper systems in place to minimize cross-contamination during sample preparation and laboratory testing?	In room 222, samples from more than one product are stored together in one small storage area. Stacking of samples in containers is required.		Lack of sufficient storage facilities and space.	Design space and storage facilities to eliminate common storage.
4.1.3 Are all controlled temperature/humidity storage areas, incubators, etc., monitored to assure that proper conditions are maintained?	There are no chart recorders installed in the QO incoming and in-process laboratory facilities to assure that proper conditions are maintained in each individual room.	No	See 4.1.7	See 4.1.7
4.1.4 Have temperature monitoring systems and equipment been properly validated?	Computer system for building management has not been validated.	No	See 4.1.7	See 4.1.7

Laboratory Control System Subelement:	*5.0 Methods Validation and Technology Transfer (MV)*
Auditor(s):	J. Foosball, W. Link, E. Vazquez, D. Blistex

Description of the QMS Subelement 5.0 Methods Validation and Technology Transfer (MV)

The Laboratory Methods Validation and Technology Transfer subelement has three individual topics as defined in the audit checklist. These are: Validation of Analytical Methods—General, Cleaning Methods Validation, and Procedures for Methods Transfer. These three topics are addressed separately as part of the subelement discussion in the following sections.

CURRENT PRACTICE 5.1 VALIDATION OF ANALYTICAL METHODS-GENERAL

This review was limited in scope due to the fact that the majority of the methods validations are initiated and performed primarily by the Research and Development Group in Tahiti. However, technology transfer is an important part of the Quality Operations Laboratory involvement with respect to methods validation. It should be noted however, that the QO Laboratory at SITE does perform some limited methods validations on older products in an effort to upgrade the quality of the methods in order to comply with CGMPs. In order to perform a complete assessment, one older product and one newer product were selected for review. Specifically, Vanilla and Egg Salad documents related to methods validation and technology transfer (including the NDA CMC sections) were reviewed using the audit checklists and additional checklists created specifically for this portion of the audit. These additional checklists that were generated using existing Level II and Level III documents for methods validation and cleaning validation. The overall results of these assessments concluded that SITE has not yet been involved in an analytical methods validation/technology transfer exercise that fully uses the guidance spelled out in the Level II and Level III documents. Therefore, it is difficult to state what the true current state of compliance with industry standards. Because of this, the Tahiti R &D Group should be consulted to determine when the next methods transfer will occur and a future audit should be scheduled for some time following its transfer.

Current Practice 5.2 Cleaning Methods Validation

This review looked at the cleaning validation documents associated with Zoofoot. A checklist which was developed using LEVEL II 22,169 *Cleaning Validation for Drug Products and Active Pharmaceutical Ingredients* in addition to the audit checklists. The cleaning validation package for Zoofoot was then reviewed against it. Cleaning validation studies are initiated by protocol at SITE and involve determining recoveries from various surfaces. Although a Level II document does exist there is no corresponding

Level III procedure. Cleaning validation summary reports are generated as the finished product for these studies.

Current Practice 5.3 Procedures for Methods Transfer

As stated in 5.1, the SITE has not yet been involved in an analytical methods validation/technology transfer exercise, which fully uses the guidance spelled out in the new Level II and Level III documents. Therefore, it is difficult to state what the true current state of compliance with industry standards is. However, methods transferred to this point have been executed via issuance of a protocol and completed by publication of technology transfer summary reports in alignment with current industry practice.

Site Documents Reviewed

- YLP-01-019
- *Zoofoot Tablets 10 mg Analytical Technology Transfer Protocol Addendum,* Nov 32, 2007
- *Zoofoot Tablets 10 mg Analytical Technology Transfer Protocol for Drug Products Methods,* MV 01-103 Nov 2. 2011
- *Zoofoot Tablets 10 mg Analytical Technology Transfer Report for Drug Products Methods,* MV 01-103 Dec 18, 2011
- *Training Program for the Analytical Laboratory Testing of Zoofoot Tablets, 10 mg,* Nov 2, 2001.
- YLP-02-019
- Master list, process validation activities for 2009
- NDA CMC Section for Egg Salad
- NDA CMC Section for Vanilla
- Protocol*: Validation of Analytical Methodology, Rinse and Swab Sampling Technique for Zoofoot API on Product Contact Surfaces for Cleaning Validation Studies*
- STP 689
- STP 690
- Summary Report*: Validation of Analytical Methodology, Rinse and Swab Sampling Technique for Zoofoot on Product Contact Surfaces for Cleaning Validation Studies* MV-00-0009
- LEVEL II 23,102
- LEVEL II 22,305
- LEVEL II 23,706

Gaps in the System versus the Audit Checklist

The following matrix correlates potential gaps uncovered during the self audit and links them to specific line times in the audit checklist. In addition to the gap, the matrix also indicates if the gap represents part of the system that is in sustainable compliance (e.g., No = a Critical finding which potentially could result in a Form 483 finding, if discovered by FDA), what the potential root cause may be for the gap, and some suggestions for potential corrective action to make the system become compliant. If the auditor did not make suggestions as to the root cause or potential corrective action to become compliant, the statement "None offered" in included in the space.

In addition to the findings correlated directly to the audit checklist, additional gaps are included in the matrix. In some circumstances the description serves as the gap and therefore the gap block may state "Same as description."

Gaps in the System: Laboratory Subelement 5.0 Methods Validation and Technology Transfer (MV)

Checklist Item Number and Description	Gap	In Substantial Compliance?	Potential Root Cause	Potential Corrective Action
5.1.1 Is there a general SOP for methods validation?	SITE is performing methods validation but does not have an active Level III SOP.	No	The SITE QC Laboratory does not have the primary responsibility for developing and validating analytical methods. However, due to the flexible nature of work they are some times called upon to perform limited methods validations/revalidations but do not have a Level III SOP to support this effort.	Create a Level III SOP based on existing Level II standard.
5.2.2 Is cleaning validation conducted according to a master plan or schedule?	SITE is not performing cleaning validation according to a master plan or schedule.		A lack of detailed scheduling and planning with Nation Wide and Local Technical Services often causes work to be performed in a rushed manner and increases excessive	Communication with Nation Wide Technical Services and Local Technical Services needs to be improved. The QO Laboratory needs to approach improving communications from a process diagram format.

Description		Root Cause	Recommendation
peak workloads. Work is often a surprise to the analysts. In addition, no single individual within the QC Laboratory communicates and coordinates these efforts.			By showing how these other departments impact upon the QO Lab workflow, a better understanding and appreciation of what the lab has to accomplish may be imparted.
9.1 Analysts at the bench level are not involved in constructing the technology transfer protocols, thus degrading the executability of the protocols, which can lead to errors.	Same as description	None offered.	Have bench-level personnel review technology transfer protocols prior to their approval.
9.2 A lack of detailed scheduling and planning with Nation Wide and Local Technical Services often causes work to be performed in a rushed manner and creates excessive peak workloads.	Same as description.	Lack of planning and lack of project management.	Communication with Nation Wide Technical Services and Local Technical Services needs to be improved. The QO Laboratory needs to approach improving

(Continued)

187

Gaps in the System: Laboratory Subelement 5.0 Methods Validation and Technology Transfer (MV) *(Continued)*

Checklist Item Number and Description	Gap	In Substantial Compliance?	Potential Root Cause	Potential Corrective Action
Work is often a surprise to the analysts. In addition, no single individual within the QC Laboratory communicates and coordinates these efforts.				communications from a process diagram format. By showing how these other departments impact upon the QO Laboratory workflow, a better understanding and appreciation of what the lab has to accomplish may be imparted. Personnel for all departments need to receive some basic project management training.

Laboratory Control System
Subelement:

6.0 Laboratory Computer
Systems (LC)

Auditor(s):

M. Gummy, W. Link,
E. Vazquez, R. Gillen

Description of the QMS Subelement 6.0 Laboratory Computer Systems (LC)

The Laboratory Computer Systems subelement has seven individual topics as defined in the audit checklist. These are: Laboratory Computer Systems—General, Centralized and Network-Attached Data Systems, Stand-Alone Data Systems, SOPs and Records, Laboratory Information Management Systems (LIMS), Spreadsheets, and Other Systems. Each of these topics is addressed separately as part of the subelement discussion in the following sections. Audit of this subelement involved interviews with laboratory representatives who are working with the site computer personnel to address computer related issues identified during previous corporate QA audits. In addition, a rewritten version of the audit checklist was developed and used to evaluate the laboratories current status.

Current Practice 6.1 Laboratory Computer Systems-General

In general, the Quality Operations Laboratory has several issues with respect to validation of computer systems and documentation, in general. As examples, there are currently no procedures in place, that govern data naming conventions for projects, analyses, products, and so on. Moreover, there currently is no disaster recovery plan in place for laboratory computer systems. These types of laboratory computer infrastructure issues impose substantial risk to the short-term and long-term security and integrity of laboratory data.

Current Practice 6.2 Centralized and Network-Attached Data Systems

The Quality Operations Laboratory is currently in the process of establishing a more formal relationship with the corporate IT group, which is located in Tahiti. As part of that relationship, SOPs are being created which address data management, incident management, performance monitoring, and so on, for server based laboratory applications. Until these SOPs are completed, no formal policy or procedure exists to address network-attached data systems.

Current Practice 6.3 Stand-Alone Data Systems

Issues related to stand-alone data systems are currently the responsibility of the individual group leaders within the Quality Operations Laboratory. These responsibilities relate to data back-up procedures and schedules, storage of back-up media, and restoration of data when necessary. None of these procedures is currently defined in a formal SOP.

Current Practice 6.4 SOPs and Records

SOPs for addressing laboratory computer systems have until recently been part of the laboratories responsibility and thus no formal SOPs exist. These responsibilities are being transferred to the corporate IT group who will generate formal, reviewed and approved SOPs. The laboratory needs to be involved in those processes.

Current Practice 6.5 Laboratory Information Management Systems (LIMS)

LIMS is currently not in use in the SITE Quality Operations Laboratory.

Current Practice 6.6 Spreadsheets

No evidence was discovered during the audit, which showed the use of unvalidated spreadsheets to generate CMGP data. Spreadsheets are currently used in the read-only mode and are unalterable. No SOP exists defining their generation and use, however.

Current Practice 6.7 Other Systems

The only system falling under this category is the electronic training record software, XTrain. This is a centrally supported application and it has been validated by the corporate IT department.

Site Documents Reviewed

- Computer server system validation protocols
- YLP 01-102
- Milan OQ/PQ software validation protocols
- LEVEL II 22,110 to LEVEL II 2,116 DRAFT

Gaps in the System versus the Audit Checklist

The following matrix correlates potential gaps uncovered during the self audit and links them to specific line times in the audit checklist. In addition to the gap, the matrix also indicates if the gap represents part of the system, that is in sustainable compliance (e.g., No = a Critical finding which potentially could result in a Form 483 finding, if discovered by FDA), what the potential root cause may be for the gap, and some suggestions for potential corrective action to make the system become compliant. If the auditors did not make suggestions as to the root cause or potential corrective action to become compliant, the statement "None offered" in included in the space.

In addition to the findings correlated directly to the audit checklist, additional gaps are included in the matrix. In some circumstances the description serves as the gap and therefore the gap block may state "Same as description."

Gaps in the System: Laboratory Subelement 6.0 Laboratory Computer Systems (LC)

Checklist Item Number and Description	Gap	In Substantial Compliance?	Potential Root Cause	Potential Corrective Action
6.1.2 Is a disaster recovery plan in place addressing laboratory data systems?	There is no IT disaster recovery plan for the site. It is being developed with cooperation of user organizations and is in draft stage at this time.	No	Lack of awareness of importance of a disaster recovery plan.	Complete development of the plan which is now in progress ensuring participation by laboratory staff.
6.1.3 And 6.1.4 Have all GMP laboratory data systems and associated file servers been validated in accordance with applicable standards, and reviewed for compliance with 21 CFR Part 11?	Laboratory data systems have not been validated in accordance with applicable standards and procedures.	No	Incomplete OQ/PQ protocols were generated during initial validation. Outdated or incomplete validations are presently in place. In addition, an inaccurate Part 11 assessment was conducted.	Investigate system status as assessed and documented by SITE, corporate (Tahiti) staff. Assess implications for SITE laboratory and recommend a course of action.
6.1.5 Are GMP workstations clearly labeled as such?	GXP workstation is not labeled.	No	Lack of training on GXP labeling.	Label correctly and retrain responsible personnel.
6.16 Is access to laboratory workstations through operating systems that provide individual	1. IT has no formal procedure which documents the	No	Lack of understanding of need for proper documentation.	Develop formal documentation of configuration and

(Continued)

191

Gaps in the System: Laboratory Subelement 6.0 Laboratory Computer Systems (LC) (*Continued*)

Checklist Item Number and Description	Gap	In Substantial Compliance?	Potential Root Cause	Potential Corrective Action
user accounts? How are they controlled?	configuration of workstations and the requirements for users who access them. 2. Windows 95 is still in use for some workstations.		Old (legacy) system	requirements for users. Provide training for all site personnel who have workstation access. Replace software application with compliant version.
6.1.7 Are workstations configured with password-controlled screensavers or other automatic locking mechanism?	One computer in immediate release area had screensaver and password turned off.		Miss configuration and/ or user modified setting.	Review all computer workstations for compliance with IT standard.
6.1.9 Are procedures in place to govern data naming conventions for projects, analyses, products, etc.?	There are currently no procedures in place, which govern data naming conventions in general.		None offered.	Implement Level II and III documents.
6.2.2 Are there documented service-level agreements with the central support organization defining mutual expectations and responsibilities?	IT has no documented, routinely updated service agreement with the lab to identify needs and expectations from systems they support.		Lack of knowledge.	Schedule a meeting to establish needs. SOP can then be written to cover the agreements.
6.2.6 Does the central support organization have server/ application SOPs which address	1. IT has an incomplete set of SOP's for basic operational functions		Lack of resources in the IT Department.	Assign priorities in order to properly develop and implement procedures.

Question		Observation	Recommendation
data management, incident management, performance monitoring, etc. specific to the laboratory?		such as configuration management, performance management, backup and restore, plus others. 2. The lab has SOP YLP 02-102 dealing with SuperSmart administration and operation, which should be transferred to IT.	By the time of implementation responsibilities were not clearly defined.	A change authorization needs to be generated to transfer procedure to IT area.
6.3.1 Are back-ups performed either automatically (scheduled through a system procedure) or manually as defined by SOP?	No	No formal SOP exists that define how data back-ups are to be performed.	None offered.	Create and implement an SOP.
6.4.1 Do SOPs exist for the validation of laboratory data systems?	No	Existing SOPs for system validation do not meet current standards.	The new Level II document on computer validation and system life-cycle needs and the supporting Level III SOPs at the site have not been completed or implemented.	Continue existing efforts on computer management.
6.6.2 Are procedures in place governing the development/ modification of spreadsheets for GMP use?		No SOP exists defining the generation and use of spreadsheets in the laboratory.	None offered.	Create and implement an SOP.

193

Laboratory Control System Subelement:	*7.0 Laboratory Investigations (LI)*
Auditor(s):	R. Minky, B. McMillan

Description of the QMS Subelement 7.0 Laboratory Investigations (LI)

The Laboratory Investigations subelement has three individual topics as defined in the audit checklist. These are: Laboratory Investigations—General, Laboratory Investigations—Execution, and Laboratory Investigations—Documentation. Each of these topics is addressed separately as part of the subelement discussion in the following sections. Audit of the subelement involved interviews with laboratory personnel who are performing investigations. In addition, a random selection of completed laboratory investigation reports (LIRs) was reviewed and assessed for their completeness and accuracy as well as the training records of all laboratory personnel associated with these investigations. Fourteen LIRs were chosen in all. The Laboratory Investigations were reviewed for:

- Description of the event
- Root cause determination
- The investigation thought process
- Conclusion
- Corrective action–preventive action recommendations
- Assessment of impact event cause investigation

Current Practice 7.1 Laboratory Investigations—General

The Quality Operations Laboratory has successfully created and implemented a Level III SOP within the last year. This SOP is based on corporate guidance documents and the Level II SOP as well. The Level III SOP is very thorough and provides sufficient detail to be used as a daily working document. All personnel engaged in performing laboratory investigations have been trained on this procedure. The SOP includes flowcharts, checklists, and forms, which structure the investigation and guide the investigator through the investigation process. Detail is included within the checklist to insure that special considerations, which may be specific for certain analytical techniques are addressed during the investigation.

Current Practice 7.2 Laboratory Investigations—Execution

Laboratory investigations are conducted by specialists who were specifically hired and trained to conduct deviation and out of specification investigations for the Quality Operations Laboratory. Currently there are two full-time investigators on staff. These investigators work closely with the analysts who generated the aberrant results, the front line supervisors, mid level management and QA to complete the investigations within the specified time frame required by SOP. The Level III SOP and associated forms are used in every case.

Current Practice 7.3 Laboratory Investigations—Documentation

The laboratory investigation process is supported by the use of a validated, in-house software program is used to insure the timely and accurate completion of the investigations. Supporting documentation is collected, compiled and filed appropriately in a paper filing system. However, all supporting paperwork is scanned and made available electronically through secure document format (pdf files). All investigations are tracked and trended as appropriate and the data are used to support annual product reviews. Whenever possible, the root causes of the events are determined and reported.

Randomly Sampled Competed Investigations—Findings

Of the 14 randomly sampled laboratory investigation reports, the overall findings were generally adequate in:

- Describing the event
- Determining the root cause
- Performing the investigation
- Supporting the conclusion with data
- Recommending a course of action or product disposition
- Providing a corrective action and instituting a preventive action
- Correctly assessing impact

Site Documents Reviewed

- Corporate Level I Guidance Document *Frame Work for Conducting Investigations*
- SITE Level II Guidance Document *Conduct Manufacturing and Laboratory Investigations*
- SOP YLP 02-007 *How to Conduct Laboratory Investigations*
- LIR 02-SUX-044
- LIR 02- SUX -069
- LIR 02- SUX -085
- LIR 02- SUX -087
- LIR 02- SUX -132
- LIR 02- SUX -144
- LIR 02- SUX -057
- LIR 02- SUX -011
- LIR 02- SUX -014
- LIR 02- SUX -017
- LIR 02- SUX -033
- LIR 02- SUX -047
- LIR 02- SUX -055
- LIR 02- SUX -066

Gaps in the System

The following matrix summarizes the gaps and indicates if the gap is considered critical (e.g., potentially could result in a Form 483 finding, if discovered by FDA), what the potential root cause may be for the gap, and some suggestions for potential corrective action to become compliant are in this case shown in the preceding narrative. Since the Quality Operations Laboratory has, within the last year, undergone a wholesale revision of their laboratory investigation system, few gaps exist for this particular Laboratory Control System subelement.

Gaps in the System: Laboratory Subelement 7.0 Laboratory Investigations (LI)

Checklist Item Number and Description	Gap	In Substantial Compliance?	Potential Root Cause	Potential Corrective Action
7.1.11 Are all laboratory personnel trained in the proper execution of laboratory investigations?	Some of the laboratory personnel responsible for creating aberrant results are not properly trained on the laboratory investigations SOP.		Since the Quality Operations Laboratory has two specialists who were specifically hired and trained to conduct deviation and out-of-specification (OOS) investigations some bench chemists have lapsed in their knowledge of the investigations process.	Insure that periodic retraining occurs on the investigations SOP for everyone in the laboratory.
9.1 A scientifically valid root cause was not determined for investigation LIR 02-SUX-033	Same as description.		The investigation specialist was not substantially familiar with the analytical technique used, which resulted in the aberrant result. Subsequently, management did not catch the lack of adequate root cause assignment.	Train the specialist on the analytical technique. Train managers to be aware that just because they have a specialist conducting investigations, they must remain actively involved in the investigation review process. The specialist must be trained to resist the urge to rubber stamp a completed investigation.

APPENDIX A SELF-AUDIT WORKFLOW

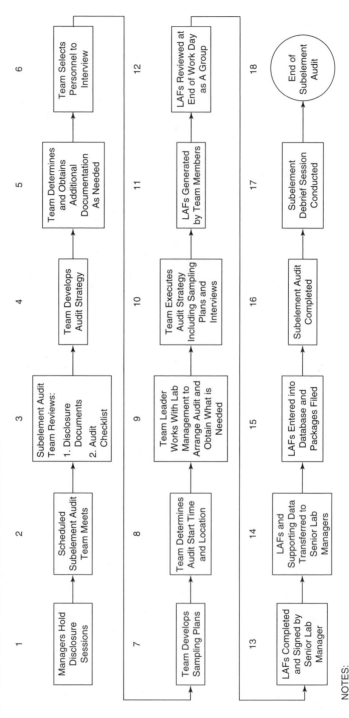

1 — Managers Hold Disclosure Sessions

2 — Scheduled Subelement Audit Team Meets

3 — Subelement Audit Team Reviews:
1. Disclosure Documents
2. Audit Checklist

4 — Team Develops Audit Strategy

5 — Team Determines and Obtains Additional Documentation As Needed

6 — Team Selects Personnel to Interview

7 — Team Develops Sampling Plans

8 — Team Determines Audit Start Time and Location

9 — Team Leader Works With Lab Management to Arrange Audit and Obtain What is Needed

10 — Team Executes Audit Strategy Including Sampling Plans and Interviews

11 — LAFs Generated by Team Members

12 — LAFs Reviewed at End of Work Day as A Group

13 — LAFs Completed and Signed by Senior Lab Manager

14 — LAFs and Supporting Data Transferred to Senior Lab Managers

15 — LAFs Entered into Database and Packages Filed

16 — Subelement Audit Completed

17 — Subelement Debrief Session Conducted

18 — End of Subelement Audit

NOTES:

APPENDIX B LAF-TO-CORRECTIVE ACTION PLAN WORKFLOW

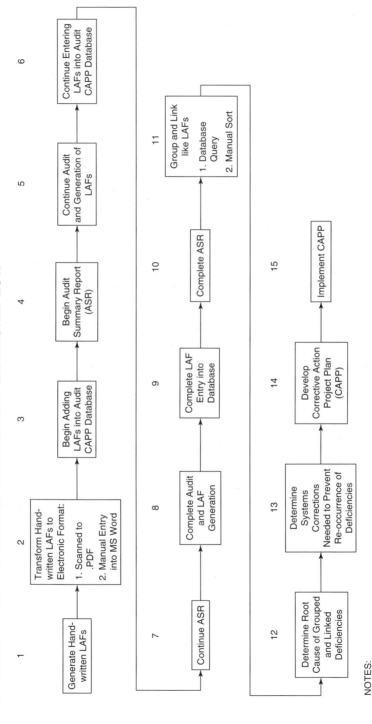

NOTES:

GLOSSARY OF CGMP AND AUDIT SYSTEM TERMS

The following terms are commonly encountered while working in a CGMP Laboratory.

483 (Form 483)   The designation of the FDA form used to capture and report observations of CGMP deficiencies related to FDA onsite audits. These observations are compiled and become part of the facilities inspection report. Form 483 observations become part of the public record.

Acceptance Criteria   Numerical limits, ranges, or other suitable measures used to determine the acceptability of the results of analytical procedures.

Accuracy   Expresses the closeness of agreement between the value found and the value that is accepted as either a conventional true value or an accepted reference value. It may often be expressed as the recovery by the assay of known, added amounts of analyte.

Action Level/Alert Level   Alert level is used to identify the point at which a parameter has drifted toward the extreme of the specified operating range. Action level is when the parameter has drifted outside of the specified operating range. Alert and action levels must be tighter than registration specifications. Alerts are reported to management and evaluated. If an action level is reached, it is reported to management, investigated and a corrective action initiated.

Establishing a CGMP Laboratory Audit System. By David M. Bliesner
Copyright © 2006 John Wiley & Sons, Inc.

Active Pharmaceutical Ingredient (API) Also known as drug substance. Any component that is intended to furnish pharmacological activity or other direct effect in the diagnosis, cure, mitigation, treatment or prevention of disease, or to affect the structure of any function of the body of man or other animals.

Analytical Performance Characteristics A term used by the USP, analytical performance characteristics refers to those characteristics of an analytical method which define its performance as an analytical technique. These performance characteristics include accuracy, precision, specificity, detection limit, quantitation limit, linearity, and range. These characteristics need to be considered when validating anyone of the USP method categories.

Atypical Result Results generated on a test article that are within specifications but are inconsistent with previous data, established trends, or other results for the same sample on test.

API See Active pharmaceutical ingredient.

Audit Summary Report (ASR) The final output of the laboratory audit. A coherent and organized presentation of findings and suggestions for corrective and preventive actions.

Batch A specific quantity of a drug or other material that is intended to have a uniform character and quality within specified limits and is produced according to a single manufacturing order during the same one cycle of manufacture.

Batch Record A record prepared for each batch of drug product or API produced that includes complete information relating to the production and control of each batch.

Blank A sample or standard of a particular matrix or composition without analyte.

Calibration The demonstration that a particular instrument or device produces results within specified limits by comparison with those produced by a reference or traceable standard over an appropriate range of measurements.

Change Control Procedure A procedure describing measures to be taken for the purpose of controlling and maintaining an audit trail when changes are made to any part of a system (e.g., standard operating procedure, test method, or specification).

Calibration Curve A calibration curve is a plot of standard solution concentration on the x axis versus instrument response on the y axis. In chromatographic analyses, calibration curves are generated by analyzing standard analyte solutions of known concentration and measuring the resulting chromatographic peak area. The resulting plot is then used to determine the concentration of unknown sample solutions containing the same analyte. This is done by measuring the unknown peak area (y) and using the equation for the line to solve for the concentration of the unknown (x). Although referred to as a curve, it is usually a linear plot with a well-defined slope and y intercept.

Capacity Factor k′   A dimensionless quantity used to describe the retention of a compound. It is calculated by the following formula: $t_r - t_0/t_0$, where t_r is the measured retention time of the component of interest and t_0 is the retention time of an unretained component. t_0 is most measured at the first disturbance of the baseline in HPLC analyses. t_r is measured at the peak apex. $k′$ is a normalized value for retention. Values range between 2 and 10 for acceptable chromatography.

CBE 30 (Change By Effect, 30 Days)   Supplemental changes to applications that do not require prior approval by FDA. These changes may be implemented within 30 days following submission to FDA if the agency has no comments.

Check Standard   A second preparation of the working standard analyzed as part of the system suitability run. The check standard is prepared at the same concentration as the working standard. Prior to continuing the chromatographic run, the ratios of the response factors (response factor = area/concentration) for the working standard and the check standard is calculated. $RF_{check\ standard}/RF_{working\ standard}$ should normally be within $\pm$ 2.0%. This provides assurance that the working standard was prepared correctly.

Code of Federal Regulations (CFR)   The Code of Federal Regulations (CFR) is the codification of the general and permanent rules published in the Federal Register by the executive departments and agencies of the federal government. It is divided into 50 titles that represent broad areas subject to federal regulation.

Compendial Tests Methods   Test methods that appear in official compendia such as the United States Pharmacopoeia (USP/NF).

Complaint   Any verbal, written, or electronic report that alleges deficiencies related to the identity, strength, quality, purity, or effectiveness of a product after it has been released for distribution.

Component   Any ingredient intended for use in the manufacture of a drug product including those that may not appear in such drug product and primary packaging components.

Compounding   Bringing together into homogenous mix of active ingredients, excipients, and as applicable, solvent components.

Consent Decree   A voluntary legal agreement a drug firm enters into with the FDA for the expressed purpose of correcting deficiencies related to CGMPs.

Contract Research Organization (CRO)   A contract manufacturing or analytical testing company.

Corrective Action Project Plan (CAPP)   The project plan generated to insure successful implementation of corrective and preventive actions (*CAPA*).

Corrective and Preventive Actions (CAPA)   The steps taken to correct and prevent deficiencies uncovered during a laboratory audit.

Current Good Manufacturing Practices (CGMPs)   21 Code of Federal Regulations Parts 210 and 211. Federal regulations that describe the minimum current good manufacturing practices for preparation of drug products for administration to human and animals. They include methods to be used in and the facilities or controls to be used for the manufacturing, processing, packing, or holding of a drug to assure that such drug meets the requirements of the act as to safety and has the identity and strength and meets the quality and purity characteristics that it purports or is represented to possess.

Degradation Product   A molecule resulting from a chemical change in the drug molecule brought about over time and/or the action of light, temperature, pH, water, etc., or by reaction with and excipient and/or the immediate container/closure system.

Detection Limit   The detection limit (DL) or limit of detection (LOD) of an individual procedure is the lowest amount of analyte in a sample that can be detected but not necessarily quantitated as an exact value. The LOD is a parameter of limit tests (tests that only determine if the analyte concentration is above or below a specification limit). In analytical procedures that exhibit baseline noise, the LOD can be based on a signal-to-noise ratio (3:1) usually expressed as the concentration (e.g., percentage, parts per billion) of analyte in the sample.

Development Report   A report that summarizes the major stages of drug product or API development from early stages through large scale manufacturing.

Document Control System   A system for managing preparation, review, approval, issuance, distribution, revision, retention, archival, obsolescence, and destruction of lifecycle documents.

Documentation   Any combination of text, graphics, data, audio, or video information that can be used to clearly and completely recreate an activity, event, or process. Documentation includes lifecycle documents as well as records.

Drug Product   The combination of API and excipients that are processed into a dosage form and marketed to the public. Common examples include tablets, capsules, oral solutions, etc. Also referred to as finished product or dosage form.

Drug Substance   See Active Pharmaceutical Ingredient (API).

Effective Date   Date by which the approved standard or procedure shall be implemented and in use. All required training must be completed prior to this date.

Excipient(s)   A raw material that may perform a variety of roles in a drug product (e.g., tablet press lubricant, filler, diluent, disintegration accelerator, colorant, etc.) However, unlike the API, which is pharmacologically active, the excipient has no intrinsic pharmacological activity.

Expiration Date   The date placed on the container/labels of a drug product or API designating the time during which a batch of product is expected to remain within the approved shelf-life specification if stored under defined conditions and after which it must not be used.

Extraction Efficiency   A measure of the effectiveness of extraction of the drug substance from the sample matrix. Studies are conducted during methods validation to determine that the sample preparation scheme is sufficient to ensure complete extraction without being unnecessarily excessive. This is normally investigated by varying the shaking or sonication times (and/or temperature) as appropriate.

Filter Study   A comparison of filter to unfiltered solutions in a methods validation to determine whether the filter being used retains any active compounds or contributes unknown compounds to the analysis.

Food and Drug Administration (FDA)   The FDA is responsible for protecting the public health by assuring the safety, efficacy, and security of human and veterinary drugs, biological products, medical devices, U.S. food supply, cosmetics, and products that emit radiation. The FDA is also responsible for advancing the public health by helping to speed innovations that make medicines and foods more effective, safer, and more affordable and securing public access to accurate, science-based information needed to use medicines and foods to improve their health.

Forced Degradation Studies   Studies undertaken to degrade the sample (e.g., drug product or API) deliberately. These studies, which may be undertaken during method development and/or validation, are used to evaluate an analytical method's ability to measure an active ingredient and its degradation products without interference. They are an integral part of the validation of a method as being specific and stability indicating.

Formulation   The recipe that describes the quantity and identity of API and excipients which make up a drug product. For example, a 100 mg tablet of Advil® may actually weigh 165 mg in total. The formulation may include: 100 mg of ibuprofen (the active ingredient), 50 mg of starch (filler), 5 mg of talc (lubricant for the tablet press), and 10 mg of iron oxide (colorant).

Good Documentation Practices   The handling of written or pictorial information describing, defining, specifying, and/or reporting of certifying activities, requirements, procedures, or results in such a way as to ensure data integrity.

ICH   See the Tripartite International Conference on Harmonization.

Identification Test   An analytical method that is capable of determining the presence of the analyte and can discriminate between closely related compounds.

In-Process Control/Test   Checks performed during production in order to monitor and, if appropriate, adjust the process and/or to ensure that the product meets its specifications. The control of the environment or equipment may also be regarded as part of in-process control.

Installation Operation and Performance Qualification (IQ/OQ/PQ)   The process by which laboratory equipment is properly installed, and determined to be operating within specifications for its intended use. IQ/OQ/PQ is executed via protocol with predetermined acceptance criteria.

Intermediate   Any substance, whether isolated or not, that is produced by chemical, physical, or biological action at some stage in the production of a drug product or API and that is subsequently used at another stage of production.

Internal Audit   A systematic examination conducted by an internal organizational unit to determine whether quality activities and related results comply with policies, standards, and requirements and whether practices have been implemented effectively and are suitable to achieve objectives and ensure compliance with regulatory requirements.

Label Claim   Theoretical strength of the product as given on the marketed product label.

Laboratory Audit Form (LAF)   The primary data capture instrument used when conducting a laboratory Audit.

Laboratory Information Management System (LIMS)   Any computer system which is used to collect, compile, organize, and report laboratory data. LIMs may also be used to calculate results as well.

Laboratory Investigation Report (LIR)   An investigation of any laboratory results or observation that does not meet acceptance criteria or falls outside the expected operational parameters. It is similar to an OOS investigation, but does not have to do with meeting a release specification.

Laboratory Qualification   A process where a laboratory that has demonstrated that it has the systems in place necessary to properly perform the tests being conducted.

Linearity   Evaluates the analytical procedure's ability (within a given range) to obtain a response that is directly proportional to the concentration (amount) of analyte in the sample. If the method is linear, the test results are directly, or by well-defined mathematical transformation, proportional to the concentration of analyte in samples within a given range. Linearity is usually expressed as the confidence limit around the slope of the regression line. The line is calculated according to an established mathematical relationship from the test response obtained by the analysis of samples with varying concentrations of analyte. Linearity may be established for all active substances, preservatives and expected impurities. Evaluation is performed on standards. Note that this is different than *range* (sometimes referred to as *linearity of method*) that is evaluated using samples and must encompass the specification range of the component assayed in the drug product.

Lot   A batch or any portion of a batch having uniform character and quality within specified limits, or in the case of a drug product, a continuous process. It is a specific identified amount produced in a unit of time or quantity in a manner that assures uniform character and quality, within specified limits.

Master Production Record   A document that specifies complete manufacturing and control instructions, sampling, and testing procedures, specifications, special notations, and precautions to be followed in production of the product and assures uniformity from batch to batch by specifying the batch size, components and quantities, and theoretical yields.

Matrix (Sample Matrix)   The components and physical form with which the analyte of interest are intimately associated. In the case of drug product, the matrix is the combination of excipients in which the active ingredient is diluted and formed within. For example, the matrix of a transdermal patch is an adhesive in which the drug substance is dissolved and fixed to a plastic backing covered by a release liner. The chemical composition and physical structure of the matrix can have a substantial effect on sample preparation and extraction of the active moiety.

Maximum Allowable Residue   The maximum allowable residue is used to calculate the acceptance criteria for cleaning verification methods. Residue limits should be practical, achievable, verifiable, and based on the most deleterious residue. Limits may be established based on the minimum known pharmacological, toxicological, or physiological activity of the API or its most deleterious component.

Method Development   The process by which methods are developed and evaluated for suitability of use as test methods and as precursors to validation.

Method Qualification   Preliminary method validation conducted during Phases 1, 2, and 3 to support the drug development process and the associated release of clinical trial material.

Method Transfer   Moving any analytical technique (chemical or microbiological) from one site area to another.

Methods Validation   A series of systematic laboratory studies where the performance characteristics of analytical procedures are established to meet the requirements for intended analytical applications. The FDA states in its guidance document that "Methods validation is the process of demonstrating that analytical procedures are suitable for their intended use. The methods validation process for analytical procedures begins with the planned and systematic collection by the applicant of the validation data to support analytical procedures."

Method Verification   The process by which compendial methods are determined to be suitable for analysis of a given test article by a given laboratory. As cited in 21 CFR Part 211.194(a) (2), method verification is used for USP/NF compendial methods or those compendia (and other recognized references such as the *AOAC Book of Methods*) that have documented evidence that the methods have been validated. At a minimum, this involves sample solution stability, specificity, and intermediate precision.

OOS Investigation   The systematic and planned search for the root cause that generated an out-of-specification result. OOS investigation should include formal reporting and a description of corrective and preventative actions taken.

(Out of Specification Investigation (OOS Result))   An examination, measurement, or test outcome that does not comply with the specification or predetermined acceptance criteria.

Preapproval Inspection (PAI)   An inspection by FDA to confirm the CGMP compliance of a drug manufacturing facility. This takes place prior to the FDA's market approval of the drug for sale.

Percent Relative Standard Deviation (Percent RSD or % RSD)   A common expression and measure of the relative precision of an analytical method for a given set of measurements. Percent RSD is calculated by dividing the standard deviation for a series of measurements by the mean of the same sets of measurements and multiplying by 100. % RSD = $(\sigma_{n-1}/\text{mean})*100$. Large % RSDs for a series of measurements indicates significant scatter and lack of precision in the technique.

Personnel Qualification   The combination of education, training, and/or experience that enables an individual to perform assigned tasks.

Process Impurity   Any component of the drug product resulting from the manufacturing process that is not the chemical entity defined as the drug substance or an excipient in the drug product.

Placebo   A formulation containing all ingredients of a drug product except the active ingredient for which the method is being developed.

Precision   An expression of the closeness of agreement (degree of scatter) between a series of measurements obtained from multiple aliquots of a homogenous sample under the prescribed conditions. The precision of an analytical procedure is usually expressed as the % RSD. According to FDA and ICH, precision may be considered at three levels, namely:

Repeatability   Refers to the use of the analytical procedure within a laboratory by a single analyst, on a single instrument, under the same operation conditions, over a short interval of time. This is sometimes referred to as *method precision.*

Intermediate Precision   Refers to variations within a laboratory as with different days, with different instruments, by different analysts, and so forth. It is formally known as *ruggedness.*

Reproducibility   The measure of the capacity of the method to remain unaffected by a variety of conditions such as different laboratories, analysts, instruments, reagent lots, elapsed assay times, days, and so on. More succinctly, the use of the method in different laboratories. Reproducibility is not part of the expected methods validation process. It is addressed during technical transfer of the method to different sites.

Prospective Validation   Validation conducted prior to the distribution of either a new product or product made under a revised manufacturing process, where the revisions may affect the product characteristics.

Protocol   An approved documented experimental design that, when executed, will demonstrate the ability of the subject method to perform as intended. Formal methods validation studies require a protocol. The protocol must have preassigned and approved acceptance criteria for each stage of the validation. The protocol may be a stand-alone document or may reference the methods validation SOP for specific details.

QA   See Quality Assurance Unit.

Q   The amount of dissolved active ingredient specified in the monograph, expressed as a percentage of the labeled content of the dosage form, and obtained during dissolution testing. For example, a tablet may have a specification stating that 85% of the active ingredient must be dissolved in 30 minutes of dissolution testing using USP apparatus II (paddles). In this case $Q = 85\%$. Also, a nefarious character in the "*Star Trek: The Next Generation*" TV series.

Qualification   Action of proving and documenting that equipment or ancillary systems are properly installed, work correctly, and actually lead to the expected results. Qualification is part of validation, but the individual steps alone do not constitute process validation.

Quality Assurance Unit   The quality assurance unit serves the role of the quality control unit as defined in 21 CFR 211.11. In this document, only the compliance function of the quality unit is addressed, not the testing functions. In the past there has been some confusion with respect to quality assurance and quality control. CGMPs now generally recognize QA = compliance and QC = testing.

Quality Control (QC)   The unit responsible for performing testing API and drug product. Often referred to as the "QC laboratory."

Quality System   A group of interrelated activities representing an integrated approach to philosophy and practices of manufacturing APIs and drug products to assure safety, identity, strength, purity, and quality.

Quality Unit   An organizational unit independent of production that ensures that the manufacture, testing, storage and distribution of drug products, active pharmaceutical ingredients, and components are performed in compliance with regulatory requirements and conformance to company policies and industry practices. Also referred to as quality assurance or QA.

Quantitation Limit   The quantitation limit (QL) or limit of quantitation (LOQ) of an individual analytical procedure is the lowest amount of analyte in a sample that can be quantitatively determined with suitable precision and accuracy. The quantitation limit is a parameter of quantitative assays for low concentrations of compounds in sample matrices and is used particularly for the determination of impurities and/or degradation products. It is usually expressed as the concentration (e.g., percentage, parts per million, etc.) of analyte in the sample. For analytical procedures that exhibit baseline noise the LOQ is generally estimated from a determination of signal-to-noise ratio (10:1) and may be confirmed by experiments.

Quarantine   The status of materials isolated physically or by other effective means to preclude their use pending a decision on their subsequent disposition.

Range   The interval between the upper and lower concentrations (amounts) of analyte in the sample (including these concentrations) for which it has been demonstrated that the analytical procedure has a suitable level of precision, accuracy, and linearity. Range is normally expressed in the same units as test results (e.g., percent, parts per million, etc.) obtained by the analytical method. Range (sometimes referred to as *linearity of method*) is evaluated using samples and must encompass the specification range of the component assayed in the drug product.

Raw Data   Raw data are defined as the original record of measurement or observation. Raw data may include, but are not limited to, printed instrument output, electronic signal output, computer output, hand recorded numbers, digital images, handdrawn diagrams, and so on. Raw data are proof of the original measurement or observation and by definition cannot be regenerated once collected.

Raw Material   Any ingredient intended for use in the production of intermediates, APIs, or drug products.

Reference Standard   A reference standard is a highly purified compound that is well characterized. It is used as a reference material to confirm the presence and/or amount of the analyte in samples.

Related Compounds   Related compounds are categorized as process impurities, degradants, or contaminants which may be found in finished drug products.

Relative Response Factor (RRF)   The ratio of the response factor of a major component to the response factor of a minor peak. This allows the accurate determination of a minor component without the need for actual standards.

Relative Retention Time (RRT)   The normalization of minor peaks to the parent peak in a chromatogram. The RRT of the parent is 1.0. Peaks eluting before the parent have RRTs <1.0. Peaks eluting after the parent have RRTs >1.0.

Repeatability   The variation experienced by a single analyst on a single instrument. Repeatability does not distinguish between variation from the instrument or system alone and from the sample preparation process.

Reporting Limit   The level, at or above the LOQ, below which values are not reported (e.g., reported as <0.05% for a reporting limit = 0.05%). The reporting limit may be defined by ICH thresholds.

Reprocessing   Introducing a previously processed material that does not conform to standards or specifications, back into the process and repeating a step or steps that are part of the established manufacturing process.

Resolution   A measure of the efficiency of the separation of two component mixtures. In chromatographic analyses, a resolution of >1.5 means two peaks are separated from each other all the way to the baseline, which is desirable.

Response Factor   A measuring of the signal generated by a detector normalized to the amount of analyte present. In HPLC it is usually the peak area for a given component (the response) divided by the concentration or mass which generated that response.

Retain Sample   Also known as reserve sample, it is an appropriately identified sample that is representative of each lot or batch of drug product or API stored under conditions consistent with the product labeling, in the same container–closure system in which the product is marketed or in one that is essentially the same characteristics. Used for testing that may be involved in analyzing complaint samples.

Retrospective Validation   Validation of a process for a product already in distribution based upon accumulated production, testing, and control data.

Revalidation   The process of partially or completely validating a method or process after changes or modifications have been made to the manufacturing process, analytical methodology, equipment, instrumentation, or other parameter(s) that may affect the quality and composition of the finished product. The USP specifically sites changes in synthesis of drug substance, changes in the composition of the drug product, and changes in the analytical procedure.

Robustness   The measure of the ability of an analytical method to remain unaffected by small but deliberate variations in method parameters (e.g., pH, mobile phase composition, temperature, instrument settings, etc.) and provide an indication of its reliability during normal usage. Robustness testing is a systematic process of varying a parameter and measuring the effect on the method by monitoring system suitability and/or the analysis of samples. It is part of the formal methods validation process.

Ruggedness   (See Intermediate Precision.) A dated term now commonly accepted as intermediate precision.

Selectivity   Selectivity is the ability of the method to separate the analyte from other components that may be present in the sample including impurities. Determination of selectivity normally includes analyzing placebo, blank, media for dissolution, dilution solvent, and mobile phase injections. Also, no chromatographic peaks, such as related compounds, should interfere with the analyte peak or internal standard peak, if applicable. Note that Selective = separate and shows every component in the sample.

Signal-to-Noise   In chromatography, the measure of average baseline noise (e.g., peak-to-peak) to the signal given by an analyte peak. S/N calculations are performed when determining LOD and LOQ.

Specification   The quality standards (e.g., tests, analytical procedures, and acceptance criteria) provided in an approved application to confirm the quality of the drug substances, drug products, intermediates, raw materials, reagents, and other components including in-process materials.

Spiked Placebo   Preparation of a sample to which known quantities of analyte are added to placebo material. Performed during validation to generate accurate and reproducible samples that are used to demonstrate recovery from the sample matrix.

Spiking   The addition of known amounts of a known compound to a standard, sample, or placebo typically for the purpose of confirming the performance of an analytical procedure or the calibration of an instrument.

Specificity   Specificity is the ability to assess unequivocally the analyte in the presence of components that may be expected to be present such as impurities, degradation products, and excipients. There must be inarguable data for a method to be specific. Specific = measure only the desired component without interference from other species that be present; separation is not necessarily required.

Stability Indicating Methodology   A validated quantitative analytical procedure or set of procedures that can detect the changes with time in the pertinent properties (e.g., active ingredient, preservative level, or appearance of degradation products) of the drug substance and drug product. These terms are used in the methodology:

Stability Indicating Assay   An assay that accurately measures the component of interest (the active ingredient(s) or degradation products) without interference from other degradation products, process impurities, excipients, or other potential interfering substances.

Stability Indication Profile   A set of procedures or assays that collectively detect changes with time although may not do so individually.

Standard and Sample Solution Stability   The stability of standards and samples is established under normal bench-top conditions, normal storage conditions, and sometimes in the instrument (e.g., an HPLC autosampler) to determine if special storage conditions are necessary, for instance refrigeration or protection from light. Stability is determined by comparing the response and impurity profile from aged standards or samples to that of a freshly prepared standard and to its own response from earlier time points. Note that these are short-term studies and are not intended to be part of the stability indication assessment or product stability program.

Stressed Studies   See Forced Degradation Studies.

Subject Matter Expert (SME)   An individual who is considered to be an expert on a particular subject due to a combination of education, training, and experience.

System Suitability   System suitability is the evaluation of the components of an analytical system to show that the performance of a system meets the standards required by a method. A system suitability evaluation usually contains its own set of parameters; for chromatographic assays, these may include tailing factors, resolution, and precision of standard peak areas and comparison to a confirmation standard, capacity factors, retention times, theoretical plates, and calibration curve linearity.

Tailing Factor   A measure of peak asymmetry. Peaks with a tailing factor of >2 are usually considered to be unacceptable due to difficulties in determine peak start-and-stop points, which complicates integration. Tailing peaks are an indication that the chromatographic conditions for a separation have not been properly optimized.

Technology Transfer   The transfer of a process or method including demonstration of equivalence according to predetermined criteria between the receiving and transferring sites.

Test Method   An approved, detailed procedure describing how to test a sample for a specified attribute (e.g., assay), including the amount required, instrumentation, reagents, sample preparation steps, data generation steps, and calculations use for evaluation.

The Tripartite International Conference on Harmonization (ICH)   An international organization formed to establish uniform guidance within the pharmaceutical industry. For drug development and manufacture, the ICH issues guidance, such as ICH Q2A *Text on Validation of Analytical Procedures* (March 1995), designed to instill uniformity within the industry with respect to various drug development and manufacturing issues. The ICH is composed of industry experts who work jointly and with FDA to develop these guidance documents.

Theoretical Plates   A dimensionless quantity used to express the efficiency or performance of a column under specific conditions. A decrease in theoretical plates can be an indication of HPLC column deterioration.

Transcription Accuracy Verification (TAV)   The process where the transcription of data from one location to another is confirmed by a second party. TAV is important for methods validation reports with summary results which are often transcribed and not linked to raw data.

USP <1225> Category I   One of four method categories for which validation data should be required. Category I methods include analytical methods for quantitation of major components of bulk drug substances or active ingredients (including preservatives) in finished pharmaceutical products. Category I methods are typically referred to as assays.

USP <1225> Category II   One of four method categories for which validation data should be required. Category II methods include analytical methods for determination of **impurities** in bulk drug substances or ***degradation compounds*** in finished pharmaceutical products. These methods include quantitative assays and limits tests.

USP <1225> Category III   One of four method categories for which validation data should be required. Category III methods include analytical methods for determination of performance characteristics (e.g., ***dissolution***, drug release).

USP <1225> Category IV   One of four method categories for which validation data should be required. Category IV methods include analytical methods used as ***identification tests***.

Validation   A documented program that provides a high degree of assurance that a specific process, method, or system will consistently produce a result meeting predetermined acceptance criteria.

Validation Characteristics   See Analytical Performance Characteristics.

Validation Parameters   See Analytical Performance Characteristics.

Validation Protocol   A prospective plan that when executed as intended, produces documented evidence that a process, method, or system has been properly validated.

Validation Report   A summary of experiments and results that demonstrate the method is suitable for its intended use and approved by responsible parties.

Verification   The act of reviewing, inspecting, testing, checking, or otherwise establishing and documenting whether items, processes, services, or documents conform to specified requirements.

APPENDIX IV

FDA COMPLIANCE PROGRAM GUIDANCE MANUAL 7356.002 *"DRUG MANUFACTURING INSPECTIONS"*

Compliance program guidance manuals are part of an internal FDA program designed to assist FDA employees in their regulatory duties. "FDA compliance programs provide guidance and instructions to FDA staff for obtaining information to help fulfill agency plans in the specified program area. These compliance programs neither create or confer any rights for, or on, any person and do not operate to bind FDA or the public. Alternative approaches may be used as long as said approaches satisfy the requirements of applicable statutes and regulations. These programs are intended for FDA personnel but are made available electronically to the public as they become available." A comprehensive listing of all FDA CPGMs can be found at www.fda.gov/ora/cpgm.

CPGM 7356.002 "DRUG MANUFACTURING INSPECTIONS" was formally implemented 1 February 2002 and is a continuing program. CGPM 7356.002 is designed to provide guidance to FDA inspectors with respect to evaluating through factory inspections including the collection and analysis of samples, the conditions and practices under which drugs and drug products are manufactured, packed, tested, and held. This manual describes the FDA's objectives and strategies for conducting drug manufacturing facility inspections,

Establishing a CGMP Laboratory Audit System. By David M. Bliesner
Copyright © 2006 John Wiley & Sons, Inc.

and is a informative reference for individuals responsible for laboratory CGMP compliance. A reproduction of CPGM 7356.002 is shown below, however; the official version is also available at the website shown above.

SUBJECT: DRUG MANUFACTURING INSPECTIONS	IMPLEMENTATION DATE 2/1/2002
	COMPLETION DATE Continuing

DATA REPORTING

PRODUCT CODES	PRODUCT/ASSIGNMENT CODES
All Human Drugs Industry codes: 50, 54–56, 59, 60–66	Domestic/Foreign Inspections: 56002 56002A Sterile products manufacture 56002B Repackers and relabelers 56002C Radioactive drugs 56002E Compressed medical gases 56002F Bulk pharmaceutical chemicals

FIELD REPORTING REQUIREMENTS

Forward a copy of each Establishment Inspection Report (EIR) for inspections classified as OAI due to CGMP deficiencies as part of any regulatory action recommendation submitted to HFD-300. For all inspections that result in the issuance of a Warning Letter, forward an electronic copy of each letter to the Division of Manufacturing and Product Quality, Case Management and Guidance Branch (HFD-325). An e-mail account has been established to receive copies of Warning Letters. The account e-mail address is CDERCGMPWL.

However, if information is encountered pertaining to inadequate post-approval reporting (Annual Reports, Supplements, Field Alert Reports, Adverse Drug Experience Reports, etc.) the information should be described in accordance with directions provided in those applicable compliance programs and under separate captions in the EIR. Data system information about these inspectional activities should be reported using applicable separate PAC(s). Expansion of coverage under these programs into a CGMP inspection must be reported under this compliance program.

The Districts are requested to use this revised compliance program for all GMP inspections.

PART I—BACKGROUND

A primary mission of the Food and Drug Administration is to conduct comprehensive regulatory coverage of all aspects of production and distribution of drugs and drug products to assure that such products meet the 501(a)(2)(B) requirements of the Act. FDA has developed two basic strategies:

1) evaluating through factory inspections, including the collection and analysis of associated samples, the conditions and practices under which drugs and drug products are manufactured, packed, tested and held, and

2) monitoring the quality of drugs and drug products through surveillance activities such as sampling and analyzing products in distribution.

This compliance program is designed to provide guidance for implementing the first strategy. Products from production and distribution facilities covered under this program are consistently of acceptable quality if the firm is operating in a state of control. The Drug Product Surveillance Program (CP 7356.008) provides guidance for the latter strategy.

The inspectional guidance in this program is structured to provide for efficient use of resources devoted to routine surveillance coverage, recognizing that in-depth coverage of all systems and all processes is not feasible for all firms on a biennial basis. It also provides for follow-up compliance coverage as needed.

PART II — IMPLEMENTATION

OBJECTIVES

The goal of this program's activities is to minimize consumers' exposure to adulterated drug products.

Under this program, inspections and investigations, sample collections and analyses, and regulatory or administrative follow-up are made:

1) to determine whether inspected firms are operating in compliance with applicable CGMP requirements, and if not, to provide the evidence for actions to prevent adulterated products from entering the market and as appropriate to remove adulterated products from the market, and to take action against persons responsible as appropriate;

2) to provide CGMP assessment which may be used in efficient determination of acceptability of the firm in the pre-approval review of a facility for new drug applications;

3) to provide input to firms during inspections to improve their compliance with regulations; and,

4) to continue FDA's unique expertise in drug manufacturing in determining the adequacy of CGMP requirements, Agency CGMP regulatory policy, and guidance documents.

STRATEGY

A. Biennial Inspection of Manufacturing Sites (includes repackaging, contract labs, etc.)

Drugs and drug products are manufactured using many physical operations to bring together components and containers and closures into a product that is released for distribution. Activities found in drug firms can be organized into systems that are sets of operations and related activities. Control of all systems helps to ensure the firm will produce drugs that are safe, have the identity and strength, and meet the quality and purity characteristics as intended.

Biennial inspections (every two years) conducted under this program:

1) reduce the risk that adulterated products are reaching the marketplace;
2) increase communication between the industry and the Agency;
3) provide for timely evaluation of new manufacturing operations in the firm; and,
4) provide for regular feedback from the Agency to individual firms on the continuing status of the firm's GMP compliance.

This program applies to all drug manufacturing operations.

Currently there are not enough FDA resources to audit every aspect of CGMP in every manufacturing facility during every inspection visit. Profile classes generalize inspection coverage from a small number of specific products to all the products in that class. This program establishes a systems approach to further generalize inspection coverage from a small number of profile classes to an overall evaluation of the firm. Reporting coverage for every profile class as defined in FACTS, in each biennial inspection, provides the most broadly resource-efficient approach. Biennial updating of all profile classes will allow for CGMP acceptability determinations to be made without delays resulting from revisiting the firm. This will speed the review process, in response to compressed time frames for application decisions and in response to provisions of the Food and Drug Administration Modernization Act of 1997 (FDAMA). This will allow for Pre-approval Inspections/Investigations Program inspections and Post-approval Audit Inspections Program inspections to focus on the specific issues related to a given application or the firm's ability to keep applications current.

The inspection is defined as audit coverage of 2 or more systems, with mandatory coverage of the Quality System (see system definitions below). Inspection options include different numbers of systems to be covered depending on the purpose of the inspection. Inspecting the minimum number of systems, or more systems as deemed necessary by the District, will provide the basis for an overall CGMP decision.

B. Inspection of Systems

Inspections of drug manufacturers should be made and reported using the system definitions and organization in this compliance program. Focusing on systems, rather than profile classes, will increase efficiency in conducting inspections because the systems are often applicable to multiple profile classes. One biennial inspection visit will result in a determination of acceptability/non-acceptability for all profile classes. Inspection coverage should be representative of all the profile classes manufactured by the firm. The efficiency will be realized because multiple visits to a firm will not be needed to cover all profile classes; delays in approval decisions will be avoided because up-to-date profile class information will be available at all times.

Coverage of a system should be sufficiently detailed, with specific examples selected, so that the system inspection outcome reflects the state of control in that system for every profile class. If a particular system is adequate, it should be adequate for all profile classes manufactured by the firm. For example, the way a firm handles "materials" (i.e., receipt, sampling, testing, acceptance, etc.) should be the same for all profile classes. The investigator should not have to inspect the Material System for each profile class. Likewise in the Production System, there are general requirements like SOP use, charge-in of components, equipment identification, in-process sampling and testing which can be evaluated through selection of example products in various profile classes. Under each system there may be something unique for a particular profile class: e.g., under the Materials System, the production of Water for Injection USP for use in manufacturing. Selecting unique functions within a system will be at the discretion of the lead Investigator. Any given inspection need not cover every system. See Part III.

Complete inspection of one system may necessitate further follow-up of some items within the activities of another system(s) to fully document the findings. However, this coverage does not constitute nor require complete coverage of these other systems.

C. A Scheme of Systems for the Manufacture of Drugs/Drug Products

A general scheme of systems for auditing the manufacture of drugs and drug products consists of the following:

1. *Quality System.* This system assures overall compliance with cGMPs and internal procedures and specifications. The system includes the quality control unit and all of its review and approval duties (e.g., change control, reprocessing, batch release, annual record review, validation protocols, and reports, etc.). It includes all product defect evaluations and evaluation of returned and salvaged drug products. See the CGMP regulation, 21 CFR 211 Subparts B, E, F, G, I, J, and K.

2. *Facilities and Equipment System.* This system includes the measures and activities which provide an appropriate physical environment and resources used in the production of the drugs or drug products. It includes:

   a) Buildings and facilities along with maintenance;

   b) Equipment qualifications (installation and operation); equipment calibration and preventative maintenance; and cleaning and validation of cleaning processes as appropriate. Process performance qualification will be evaluated as part of the inspection of the overall process validation which is done within the system where the process is employed; and,

   c) Utilities that are not intended to be incorporated into the product such as HVAC, compressed gases, steam and water systems.

   See the CGMP regulation, 21 CFR 211 Subparts B, C, D, and J.

3. *Materials System.* This system includes measures and activities to control finished products, components, including water or gases, that are incorporated into the product, containers and closures. It includes validation of computerized inventory control processes, drug storage, distribution controls, and records. See the CGMP regulation, 21 CFR 211 Subparts B, E, H, and J.

4. *Production System.* This system includes measures and activities to control the manufacture of drugs and drug products including batch compounding, dosage form production, in-process sampling and testing, and process validation. It also includes establishing, following, and documenting performance of approved manufacturing procedures. See the CGMP regulation, 21 CFR 211 Subparts B, F, and J.

5. *Packaging and Labeling System.* This system includes measures and activities that control the packaging and labeling of drugs and drug products. It includes written procedures, label examination and usage, label storage and issuance, packaging and labeling operations controls, and validation of these operations. See the CGMP regulation, 21 CFR 211 Subparts B, G, and J.

6. *Laboratory Control System.* This system includes measures and activities related to laboratory procedures, testing, analytical methodology development and validation or verification, and the stability program. See the CGMP regulation, 21 CFR 211 Subparts B, I, J, and K.

The overall theme in devising this scheme of systems was the subchapter structure of the CGMP regulation. Every effort was made to group whole subchapters together in a rational set of six systems which incorporates the general scheme of pharmaceutical manufacturing operations.

The organization and personnel, including appropriate qualifications and training, employed in any given system, will be evaluated as part of that system's operation. Production, control, or distribution records required to be maintained by the CGMP regulation and selected for review should be included for inspection audit within the context of each of the above systems. Inspections of contract companies should be within the system for which the product or service is contracted as well as their Quality System.

As this program approach is implemented, the experience gained will be reviewed to make modifications to the system definitions and organization as needed.

PROGRAM MANAGEMENT INSTRUCTIONS

A. Definitions

1. Surveillance Inspections

The Full Inspection Option

The Full Inspection Option is a surveillance or compliance inspection which is meant to provide a broad and deep evaluation of the firm's CGMP. This will be done when little or no information is known about a firm's CGMP compliance (e.g., for new firms); or for firms where there is doubt about the CGMP compliance in the firm (e.g., a firm whose history has documented short-lived compliance and recidivism); or follow up to previous regulatory actions. Based on findings of objectionable conditions as listed in Part V in one or more systems (a minimum of two systems must be completed), a Full Inspection may revert to the abbreviated inspection option, with District concurrence. See Part III, Section B.1. During the course of a Full Inspection, verification of quality system activities may require limited coverage in other systems. The Full Inspection Option will normally include an inspection audit of at least four of the systems, one of which must be the Quality System (the system which includes the responsibility for the annual product reviews).

The Abbreviated Inspection Option

The Abbreviated Inspection Option is a surveillance or compliance inspection which is meant to provide an efficient update evaluation of a firm's CGMP. The abbreviated inspection will provide documentation for continuing a firm in a satisfactory CGMP compliance status. Generally this will be done when a firm has a record of satisfactory CGMP compliance, with no significant recall, or product defect or alert incidents, or with little shift in the manufacturing profiles of the firm within the previous two years. See Part III, Section B.2. A Full Inspection may revert to an abbreviated inspection based on findings of objectionable conditions as listed in Part V in one or more systems. The Abbreviated Inspection Option normally will include an inspection audit of at least two of the systems, one of which must be the Quality System (the system which includes the responsibility for the annual product reviews). The District drug program managers should ensure that the optional systems are rotated in successive Abbreviated Inspections. During the course of an abbreviated inspection, verification of quality system activities may require limited coverage in other systems. Some firms participate in a limited part of the production of a drug or drug product, e.g., a contract laboratory. Such firms may employ only two of the systems defined. In these cases the inspection of the two systems will comprise inspection of the entire firm and will be considered the Full Inspection Option.

Selecting Systems for Coverage

The selection of the system(s) for coverage will be made by the District Office based on such factors as a given firm's specific operation, history of previous coverage, history of compliance, or other priorities determined by the District Office.

2. Compliance Inspections

Compliance Inspections are inspections done to evaluate or verify compliance corrective actions after a regulatory action has been taken. First, the coverage given in compliance inspections must be related to the areas found deficient and subjected to corrective actions.

In addition, coverage must be given to systems because a determination must be made on the overall compliance status of the firm after the corrective actions are taken. The firm is expected to address all of its operations in its corrective action plan after a previously violative inspection, not just the deficiencies noted in the 483. The Full Inspection Option should be used for a compliance inspection, especially if the Abbreviated Inspection Option was used during the violative inspection.

Compliance Inspections include For Cause Inspections. For Cause Inspections are compliance inspections which are done to investigate a specific problem that has come to the attention of some level of the agency. The problems may be in Field Alert Reports (FARs), industry complaints, recalls, indicators of defective products, etc. Coverage of these areas may be assigned under other compliance programs, however, expansion of the coverage to a GMP inspection is to be reported under this program. For Cause Inspections may be assigned under this program as the need arises.

3. State of Control

A drug firm is considered to be operating in a state of control when it employs conditions and practices that assure compliance with the intent of Sections 501(a)(2)(B) of the Act and portions of the CGMP regulations that pertain to their systems. A firm in a state of control produces finished drug products for which there is an adequate level of assurance of quality, strength, identity and purity.

A firm is out of control if any one system is out of control. A system is out of control if the quality, identity, strength and purity of the products resulting from that system(s) cannot be assured adequately. Documented CGMP deficiencies provide the evidence for concluding that a system is not operating in a state of control. See Part V. Regulatory/Administrative Strategy for a discussion of compliance actions based on inspection findings demonstrating out of control systems/firm.

4. Drug Process

A drug process is a related series of operations which result in the preparation of a drug or drug product. Major operations or steps in a drug process may include mixing, granulation, encapsulation, tableting, chemical synthesis, fermentation, aseptic filling, sterilization, packing, labeling, testing, etc.

5. Drug Manufacturing Inspection

A drug manufacturing inspection is a factory inspection in which evaluation of two or more systems, including the Quality System, is done to determine if manufacturing is occurring in a state of control.

B. Inspection Planning

The Field will conduct drug manufacturing inspections and maintain profiles or other monitoring systems which will ensure that each drug firm receives biennial inspectional coverage, as provided for in the strategy.

The District Office is responsible for determining the depth of coverage given to each drug firm. CGMP inspectional coverage shall be sufficient to assess the state of compliance for each firm.

The frequency and depth of inspection should be determined by the statutory obligation, the firm's compliance history, the technology employed, and the characteristics of the products. When a system is inspected, the inspection of that system may be considered applicable to all products which use it. Investigators should select an adequate number and type of products to accomplish coverage of the system. Selection of products should be made so that coverage is representative of the firms overall abilities in manufacturing within CGMP requirements.

Review of NDA/ANDA files may assist in selecting significant drug processes for coverage in the various systems. Significant drug processes are those which utilize all the systems in the firm very broadly and/or which contain steps with unique or difficult manipulation in the performance of a step. Products posing special manufacturing features, e.g., low dose products, narrow therapeutic range drugs, combination drugs, modified release products, etc., and new products made under an approved drug application, should be considered first in selecting products for coverage.

The health significance of certain CGMP deviations may be lower when the drug product involved has no major systemic effect or no dosage limitations such as in products like calamine lotion or OTC medicated shampoos. Such products should be given inspection coverage with appropriate priority.

Inspections for this compliance program may be performed during visits to a firm when operations are being performed for other compliance programs or other investigations.

C. Profiles

The inspection findings will be used as the basis for updating all profile classes in the profile screen of the FACTS EIR coversheet that is used to record profile/class determinations. Normally, an inspection under this systems approach will result in all profile classes being updated.

PART III — INSPECTIONAL

INVESTIGATIONAL OPERATIONS

A. General

Review and use the CGMPs for Finished Pharmaceuticals (21 CFR 210 and 211) to evaluate manufacturing processes. Use Guides to Inspection published by the Office of Regional Operations for information on technical applications in various manufacturing systems.

The investigator should conduct inspections according to the STRATEGY section in Part II of this compliance program. Recognizing that drug firms vary greatly in size and scope, and manufacturing systems are more or less sophisticated, the approach to inspecting each firm should be carefully planned. For example, it may be more appropriate to review the Quality

System thoroughly before entering production areas in some firms; in others, the Quality System review should take place concurrently with inspection of another system or systems selected for coverage. The complexity and variability necessitate a flexible inspection approach; one which allows the investigator to choose the inspection focus and depth appropriate for a specific firm, but also one which directs the performance and reporting on the inspection within a framework which will provide for a uniform level of CGMP assessment. Furthermore, this inspection approach will provide for fast communication and evaluation of findings.

Inspectional Observations noting CGMP deficiencies should be related to a requirement. Requirements for manufacture of drug products (dosage forms) are in the CGMP regulation and are amplified by policy in the Compliance Policy Guides, case precedents, etc. CGMP requirements apply to the manufacture of distributed prescription drug products, OTC drug products, approved products and products not requiring approval, as well as drug products used in clinical trials. The CGMP regulations are not direct requirements for manufacture of API's; the regulations should not be referenced as the basis for a GMP deficiency in the manufacture of Active Pharmaceutical Ingredients (APIs), but they are guidance for GMP in API manufacture.

Guidance documents do not establish requirements. They state examples of ways to meet requirements. Guidance documents are not to be referred to as the justification for an inspectional observation. The justification comes from the CGMPs. Current Guides to Inspection and Guidance to Industry documents provide interpretations of requirements, which may assist in the evaluation of the adequacy of CGMP systems.

Current inspectional observation policy as stated in the IOM says that the FDA-483, when issued, should be specific and contain only significant items. For this program, inspection observations should be organized under separate captions by the systems defined in this program. List observations in order of importance within each system. Where repeated or similar observations are made, they should be consolidated under a unified observation. For those Districts utilizing Turbo EIR, a limited number of observations can be common to more than one system (e.g., organization and personnel including appropriate qualifications and training.) In these instances, put the observation in the first system reported on the FDA-483 and in the text of the EIR, reference the applicability to other systems where appropriate. This is being done to accommodate the structure of Turbo EIR which allows individual citation once per FDA-483. Refrain from using unsubstantiated conclusions. Do not use the term "inadequate" without explaining why and how. Refer to policy in the IOM, Chapter 5, Section 512 and Field Management Directive 120 for further guidance on the content of Inspectional Observations.

Specific specialized inspectional guidance may be provided as attachments to this program, or in requests for inspection, assignments, etc.

B. INSPECTION APPROACHES

This program provides two surveillance inspectional options, Abbreviated Inspection Option and Full Inspection Option. See the definitions of the inspection options in Part II of this program.

1. *Selecting the Full Inspection Option.* The Full Inspection Option will include inspection of at least four of the systems as listed in Part II Strategy, one of which must be the Quality System.

   a. Select the Full Inspection Option for an initial FDA inspection of a facility. A Full Inspection may revert to the Abbreviated Inspection Option, **with District concurrence**, based on finding of objectionable conditions as listed in Part V in one or more systems (a minimum of two systems must be completed).

   b. Select the Full Inspection Option when the firm has a history of fluctuating into and out of compliance. To determine if the firm meets this criterion, the District should utilize all information at their disposal, such as, inspection results, results of sample analyses, complaints, DQRS reports, recalls, etc. and the compliance actions resulting from them or from past inspections. A Full Inspection may revert to the Abbreviated Inspection Option, **with District concurrence**, based on findings of objectionable conditions as listed in Part V in one or more systems (a minimum of two systems must be completed).

   c. Evaluate if important changes have occurred by comparing current operations against the EIR for the previous Full Inspection. The following types of changes are typical of those that warrant the Full Inspection Option:

     (1) New potential for cross-contamination arising through change in process or product line.

     (2) Use of new technology requiring new expertise, significant new equipment, or new facilities.

   d. A Full Inspection may also be conducted on a surveillance basis at the District's discretion.

   e. The Full Inspection Option will satisfy the biennial inspection requirement.

   f. Follow up to a Warning Letter or other significant regulatory actions should require a Full Inspection option.

2. *Selecting the Abbreviated Inspection Option.* The Abbreviated Inspection Option normally will include inspection audit of at least two of the above systems, one of which must be the Quality System. During the course of an abbreviated inspection, verification of quality system activities may require limited coverage in other systems.

   a. This option involves an inspection of the manufacturer to maintain surveillance over the firm's activities and to provide input to the firm on maintaining and improving the GMP level of assurance of quality of its products.

   b. A Full Inspection may revert to the Abbreviated Inspection option, **with District concurrence**, based on findings of objectionable conditions as listed in Part V in one or more systems (a minimum of two systems must be completed).

   c. An abbreviated inspection is adequate for routine coverage and will satisfy the biennial inspectional requirement.

Comprehensive Inspection Coverage

It is not anticipated that Full Inspections will be conducted every two years. They may be conducted at less frequent intervals, perhaps at every third or fourth inspection cycle. Districts should consider selecting different optional systems for inspection coverage as a cycle of Abbreviated Inspections are carried out to build comprehensive information on the firm's total manufacturing activities.

C. System Inspection Coverage

QUALITY SYSTEM

Assessment of the Quality System is two phased. The first phase is to evaluate whether the Quality Control Unit has fulfilled their responsibility to review and approve all procedures related to production, quality control, and quality assurance and assure the procedures are adequate for their intended use. This also includes the associated recordkeeping systems. The second phase is to assess the data collected to identify quality problems and may link to other major systems for inspectional coverage.

For each of the following, the firm should have written and approved procedures and documentation resulting therefrom. The firm's adherence to written procedures should be verified through observation whenever possible. These areas are not limited only to finished products, but may also incorporate starting materials and in-process materials. These areas may indicate deficiencies not only in this system but also in other major systems that would warrant expansion of coverage. All areas under this system should be covered; however the depth of coverage may vary depending upon inspectional findings.

- Product reviews: at least annually; should include information from areas listed below as appropriate; batches reviewed, for each product, are representative of all batches manufactured; trends are identified; refer to 21 CFR 211.180(e).
- Complaint reviews (quality and medical): documented; evaluated; investigated in a timely manner; includes corrective action where appropriate.
- Discrepancy and failure investigations related to manufacturing and testing: documented; evaluated; investigated in a timely manner; includes corrective action where appropriate.
- Change Control: documented; evaluated; approved; need for revalidation assessed.
- Product Improvement Projects: for marketed products
- Reprocess/Rework: evaluation, review and approval; impact on validation and stability.
- Returns/Salvages: assessment; investigation expanded where warranted; disposition.
- Rejects: investigation expanded where warranted; corrective action where appropriate.
- Stability Failures: investigation expanded where warranted; need for field alerts evaluated; disposition.
- Quarantine products.
- Validation: status of required validation/revalidation (e.g., computer, manufacturing process, laboratory methods).
- Training/qualification of employees in quality control unit functions.

FACILITIES AND EQUIPMENT SYSTEMS

For each of the following, the firm should have written and approved procedures and documentation resulting therefrom. The firm's adherence to written procedures should be verified through observation whenever possible. These areas may indicate deficiencies not only in this system but also in other systems that would warrant expansion of coverage. When this system is selected for coverage in addition to the Quality System, all areas listed below should be covered; however, the depth of coverage may vary depending upon inspectional findings.

1. Facilities
   - cleaning and maintenance
   - facility layout and air handling systems for prevention of cross-contamination (e.g., penicillin, beta-lactams, steroids, hormones, cytotoxics, etc.)
   - specifically designed areas for the manufacturing operations performed by the firm to prevent contamination or mix-ups

  - general air handling systems
  - control system for implementing changes in the building
  - lighting, potable water, washing and toilet facilities, sewage and refuse disposal
  - sanitation of the building, use of rodenticides, fungicides, insecticides, cleaning and sanitizing agents
2. Equipment
  - equipment installation and operational qualification where appropriate
  - adequacy of equipment design, size, and location
  - equipment surfaces should not be reactive, additive, or absorptive
  - appropriate use of equipment operations substances, (lubricants, coolants, refrigerants, etc.) contacting products/containers/etc.
  - cleaning procedures and cleaning validation
  - controls to prevent contamination, particularly with any pesticides or any other toxic materials, or other drug or non-drug chemicals
  - qualification, calibration and maintenance of storage equipment, such as refrigerators and freezers for ensuring that standards, raw materials, reagents, etc. are stored at the proper temperatures
  - equipment qualification, calibration and maintenance, including computer qualification/validation and security
  - control system for implementing changes in the equipment
  - equipment identification practices (where appropriate)
  - documented investigation into any unexpected discrepancy

MATERIALS SYSTEM

For each of the following, the firm should have written and approved procedures and documentation resulting therefrom. The firm's adherence to written procedures should be verified through observation whenever possible. These areas are not limited only to finished products, but may also incorporate starting materials and in-process materials. These areas may indicate deficiencies not only in this system but also in other systems that would warrant expansion of coverage. When this system is selected for coverage in addition to the Quality System, all areas listed below should be covered; however, the depth of coverage may vary depending upon inspectional findings.

  - training/qualification of personnel
  - identification of components, containers, closures
  - inventory of components, containers, closures
  - storage conditions
  - storage under quarantine until tested or examined and released
  - representative samples collected, tested or examined using appropriate means
  - at least one specific identity test is conducted on each lot of each component
  - a visual identification is conducted on each lot of containers and closures
  - testing or validation of supplier's test results for components, containers and closures
  - rejection of any component, container, closure not meeting acceptance requirements. Investigate fully the firm's procedures for verification of the source of components.

- appropriate retesting/reexamination of components, containers, closures
- first in-first out use of components, containers, closures
- quarantine of rejected materials
- water and process gas supply, design, maintenance, validation and operation
- containers and closures should not be additive, reactive, or absorptive to the drug product
- control system for implementing changes in the materials handling operations
- qualification/validation and security of computerized or automated processes
- finished product distribution records by lot
- documented investigation into any unexpected discrepancy

PRODUCTION SYSTEM

For each of the following, the firm should have written and approved procedures and documentation resulting therefrom. The firm's adherence to written procedures should be verified through observation whenever possible. These areas are not limited only to finished products, but may also incorporate in-process materials. These areas may indicate deficiencies not only in this system but also in other systems that would warrant expansion of coverage. When this system is selected for coverage in addition to the Quality System, all areas listed below should be covered; however, the depth of coverage may vary depending upon inspectional findings.

- training/qualification of personnel
- control system for implementing changes in processes
- adequate procedure and practice for charge-in of components
- formulation/manufacturing at not less than 100%
- identification of equipment with contents, and where appropriate phase of manufacturing and/or status
- validation and verification of cleaning/sterilization/depyrogenation of containers and closures
- calculation and documentation of actual yields and percentage of theoretical yields
- contemporaneous and complete batch production documentation
- established time limits for completion of phases of production
- implementation and documentation of in-process controls, tests, and examinations (e.g., pH, adequacy of mix, weight variation, clarity)
- justification and consistency of in-process specifications and drug product final specifications
- prevention of objectionable microorganisms in non-sterile drug products
- adherence to preprocessing procedures (e.g., set-up, line clearance, etc.)
- equipment cleaning and use logs
- master production and control records
- batch production and control records
- process validation, including validation and security of computerized or automated processes
- change control; the need for revalidation evaluated
- documented investigation into any unexpected discrepancy

PACKAGING AND LABELING SYSTEM

For each of the following, the firm should have written and approved procedures and documentation resulting therefrom. The firm's adherence to written procedures should be verified through observation whenever possible. These areas are not limited only to finished products, but may also incorporate starting materials and in-process materials. These areas may indicate deficiencies not only in this system but also in other systems that would warrant expansion of coverage. When this system is selected for coverage in addition to the Quality System, all areas listed below should be covered; however, the depth of coverage may vary depending upon inspectional findings.

- training/qualification of personnel
- acceptance operations for packaging and labeling materials
- control system for implementing changes in packaging and labeling operations
- adequate storage for labels and labeling, both approved and returned after issued
- control of labels which are similar in size, shape, and color for different products
- finished product cut labels for immediate containers which are similar in appearance without some type of 100 percent electronic or visual verification system or the use of dedicated lines
- gang printing of labels is not done, unless these are differentiated by size, shape, or color
- control of filled unlabeled containers that are later labeled under multiple private labels
- adequate packaging records that will include specimens of all labels used
- control of issuance of labeling, examination of issued labels and reconciliation of used labels
- examination of the labeled finished product
- adequate inspection (proofing) of incoming labeling
- use of lot numbers, destruction of excess labeling bearing lot/control numbers
- physical/spatial separation between different labeling and packaging lines
- monitoring of printing devices associated with manufacturing lines
- line clearance, inspection and documentation
- adequate expiration dates on the label
- conformance to tamper-evident (TEP) packaging requirements (see 21CFR 211.132 and Compliance Policy Guide, 7132a.17)
- validation of packaging and labeling operations including validation and security of computerized processes
- documented investigation into any unexpected discrepancy

LABORATORY CONTROL SYSTEM

For each of the following, the firm should have written and approved procedures and documentation resulting therefrom. The firm's adherence to written procedures should be verified through observation whenever possible. These areas are not limited only to finished products, but may also incorporate starting materials and in-process materials. These areas may indicate deficiencies not only in this system but also in other systems that would warrant expansion of coverage. When this system is selected for coverage in addition to the Quality System, all areas listed below should be covered; however, the depth of coverage may vary depending upon inspectional findings.

- training/qualification of personnel
- adequacy of staffing for laboratory operations
- adequacy of equipment and facility for intended use
- calibration and maintenance programs for analytical instruments and equipment
- validation and security of computerized or automated processes
- reference standards; source, purity and assay, and tests to establish equivalency to current official reference standards as appropriate
- system suitability checks on chromatographic systems (e.g., GC or HPLC)
- specifications, standards, and representative sampling plans
- adherence to the written methods of analysis
- validation/verification of analytical methods
- control system for implementing changes in laboratory operations
- required testing is performed on the correct samples
- documented investigation into any unexpected discrepancy
- complete analytical records from all tests and summaries of results
- quality and retention of raw data (e.g., chromatograms and spectra)
- correlation of result summaries to raw data; presence of unused data
- adherence to an adequate Out of Specification (OOS) procedure which includes timely completion of the investigation
- adequate reserve samples; documentation of reserve sample examination
- stability testing program, including demonstration of stability indicating capability of the test methods

D. Sampling

Samples of defective product constitute persuasive evidence that significant CGMP problems exist. Physical samples may be an integral part of a CGMP inspection where control deficiencies are observed. Physical samples should be correlated with observed control deficiencies. Consider consulting your servicing laboratory for guidance on quantity and type of samples (in-process or finished) to be collected. Documentary samples may be submitted when the documentation illustrates the deficiencies better than a physical sample. Districts may elect to collect, but not analyze, physical samples, or to collect documentary samples to document CGMP deficiencies. Physical sample analysis is not necessary to document CGMP deficiencies.

When a large number of products have been produced under deficient controls, collect physical and/or documentary samples of products which have the greatest therapeutic significance, narrow range of toxicity, or a low dosage strength. Include samples of products of minimal therapeutic significance only when they illustrate highly significant CGMP deficiencies.

For sampling guidance, refer to IOM, Chapter 4

E. Inspection Teams

An inspection team (See IOM 502.4) composed of experts from within the District, other Districts, or Headquarters is encouraged when it provides needed expertise and experience.

Contact ORO/Division of Emergency and Investigational Operations if technical assistance is needed (See also FMD 142). Participation of an analyst (chemist or microbiologist) on an inspection team is also encouraged, especially where laboratory issues are extensive or complex. Contact your Drug Servicing Laboratory or ORO/Division of Field Science.

F. Reporting

The investigator will utilize Subchapter 590 of the IOM for guidance in reporting of inspectional findings. Identify systems covered in the Summary of Findings. Identify and explain in the body of the report the rationale for inspecting the profile classes covered. Report and discuss in full any adverse findings by systems under separate captions. Add additional information as needed or desired, for example, a description of any significant changes that have occurred since previous inspections.

Reports with specific, specialized information required should be prepared as instructed within the individual assignment/attachment.

PART IV — ANALYTICAL

ANALYZING LABORATORIES

1. Routine chemical analyses — all Servicing Laboratories except WEAC.
2. Sterility testing:

Region	Examining Laboratory
NE	NRL
SE	SRL
CE	NRL
SW, PA	SAN-DO

3. Other microbiological examinations-NRL (for the CE Region), SRL, SAN, and DEN. Salmonella Serotyping Lab-ARL
4. Chemical cross-contamination analyses by mass spectrometry (MS)-NRL, SRL, DEN, PRL/NW, and PHI. Non-mass spectrometry laboratories should call one of their own regional MS capable laboratories or Division of Field Science (HFC-140) to determine the most appropriate lab for the determinations to be performed.
5. Chemical cross-contamination analyses by Nuclear Magnetic Resonance (NMR) spectroscopy-NRL. Non-NMR laboratories should call one of their own regional labs equipped with NMR or Division of Field Science (HFC-140) to determine the most appropriate lab for the determinations to be performed.
6. Dissolution testing-NRL, KAN, SRL, SJN, DET, PHI, DEN, PRL/SW and PRL-NW. Districts without dissolution testing capability should use one of their own regional labs for dissolution testing. Otherwise, call DFS.
7. Antibiotic Analyses:

ORA Examining Laboratory

Denver District Lab (HFR-SW260)
Tetracyclines
Erythromycins

Northeast Regional Lab (HFR-NE500)

Penicillins

Cephalosporins

CDER Examining Laboratory

Office of Testing and Research

Division of Pharmaceutical Analysis

(HFD-473)

All other Antibiotics

8. Bioassays—Division of Testing and Applied Analytical Research, Drug Bioanalysis Branch (HFN-471).

9. Particulate Matter in Injectables: NRL, SRL.

10. Pyrogen/LAL Testing: SRL

ANALYSIS

1. Samples are to be examined for compliance with applicable specifications as they relate to deficiencies noted during the inspection. Check analyses will be by the official method, or when no official method exists, by other validated procedures. See CPG 7152.01.

2. The presence of cross-contamination must be confirmed by a second method. Spectroscopic methods, such as MS, NMR, UV-Visible, or infrared (IR) are preferred. A second confirmatory method should be employed by different mechanisms than the initial analysis (i.e., ion-pairing vs. conventional reverse phase HPLC).

3. Check Analysis for dissolution rate must be performed by a second dissolution-testing laboratory.

4. Sterility testing methods should be based on current editions of USP, and for the Sterility Analytical Manual. Other microbiological examinations should be based on appropriate sections of USP and BAM.

PART V — REGULATORY/ADMINISTRATIVE STRATEGY

Inspection findings that demonstrate that a firm is not operating in a state of control may be used as evidence for taking appropriate advisory, administrative and/or judicial actions.

When the management of the firm is unwilling or unable to provide adequate corrective actions in an appropriate time frame, formal agency regulatory actions will be recommended, designed to meet the situation encountered.

When deciding the type of action to recommend, the initial decision should be based on the seriousness of the problem and the most effective way to protect the consumer. Outstanding instructions in the Regulatory Procedures Manual (RPM) should be followed.

The endorsement to the inspection report should point out the actions that have been taken or will be taken and when. All deficiencies noted in inspections/audits under this program must be addressed by stating the firm's corrective actions, accomplished or projected, for each as established in the discussion with management at the close of the inspection.

All corrective action approaches in domestic firms are monitored and managed by the District Offices. The approaches may range from shut down of operations, recall of products, conducting testing programs, development of new procedures, modifications of plants and equipment, to simple immediate corrections of conditions. CDER/DMPQ/CMGB/HFD-325 will assist District Offices as requested.

An inspection report that documents that one or more systems is/are out of control should be classified OAI. District Offices may issue Warning Letters per RPM, Chapter 4, to warn firms of violations, to solicit voluntary corrections, and to provide for the initial phase of formal agency regulatory actions.

Issuance of a Warning Letter or taking other regulatory actions pursuant to a surveillance inspection (other than a For Cause Inspection) should result in the classification of all profile classes as unacceptable. Also, the inspection findings will be used as the basis for updating profile classes in FACTS.

FDA laboratory tests which demonstrate effects of absent or inadequate good manufacturing practice are strong evidence for supporting regulatory actions. Such evidence development should be considered as an inspection progresses and deficiencies are found. However, the lack of violative physical samples is *not* a barrier to pursuing regulatory and/or administrative action provided that CGMP deficiencies have been well documented. Likewise, physical samples found to be in compliance are *not* a barrier to pursuing action under CGMP charges.

Evidence to support significant and/or a trend of deficiencies within a system covered could constitute the failure of a system and should result in consideration of the issuance of a Warning Letter or other regulatory actions by the District. When deciding the type of action to recommend, the initial decision should be based on the seriousness and/or the frequency of the problem. Examples include the following:

Quality System

1) Pattern of failure to review/approve procedures.
2) Pattern of failure to document execution of operations as required.
3) Pattern of failure to review documentation.
4) Pattern of failure to conduct investigations and resolve discrepancies/failures/deviations/complaints.
5) Pattern of failure to assess other systems to assure compliance with GMP and SOPs.

Facilities and Equipment

1) Contamination with filth, objectionable microorganisms, toxic chemicals or other drug chemicals, or a reasonable potential for contamination, with demonstrated avenues of contamination, such as airborne or through unclean equipment
2) Pattern of failure to validate cleaning procedures for non-dedicated equipment. Lack of demonstration of effectiveness of cleaning for dedicated equipment.
3) Pattern of failure to document investigation of discrepancies.
4) Pattern of failure to establish/follow a control system for implementing changes in the equipment.
5) Pattern of failure to qualify equipment, including computers.

Materials System

1) Release of materials for use or distribution that do not conform to established specifications.
2) Pattern of failure to conduct one specific identity test for components.
3) Pattern of failure to document investigation of discrepancies.
4) Pattern of failure to establish/follow a control system for implementing changes in the materials handling operations.
5) Lack of validation of water systems as required depending upon the intended use of the water.
6) Lack of validation of computerized processes.

Production System

1) Pattern of failure to establish/follow a control system for implementing changes in the production system operations.
2) Pattern of failure to document investigation of discrepancies
3) Lack of process validation.
4) Lack of validation of computerized processes.
5) Pattern of incomplete or missing batch production records.
6) Pattern of nonconformance to established in-process controls, tests, and/or specifications.

Packaging and Labeling

1) Pattern of failure to establish/follow a control system for implementing changes in the packaging and/or labeling operations.
2) Pattern of failure to document investigation of discrepancies.
3) Lack of validation of computerized processes.
4) Lack of control of packaging and labeling operations that may introduce a potential for mislabeling.
5) Lack of packaging validation.

Laboratory Control System

1) Pattern of failure to establish/follow a control system for implementing changes in the laboratory operations.
2) Pattern of failure to document investigation of discrepancies.
3) Lack of validation of computerized and/or automated processes.
4) Pattern of inadequate sampling practices.
5) Lack of validated analytical methods.
6) Pattern of failure to follow approved analytical procedures.
7) Pattern of failure to follow an adequate OOS procedure.
8) Pattern of failure to retain raw data.
9) Lack of stability indicating methods.
10) Pattern of failure to follow the stability programs.

Follow up to a Warning Letter or other significant regulatory actions as a result of an abbreviated inspection should warrant Full Inspection coverage as defined in this program.

PART VI — REFERENCES, ATTACHMENTS, AND PROGRAM CONTACTS

REFERENCES

1. Federal Food, Drug, and Cosmetic Act, as amended
2. Code of Federal Regulations, Title 21, Parts 210 and 211, as revised, including the General Comments (preamble)
3. Compliance Policy Guides Manual, Chapter 4 Human Drugs
4. Compressed Medical Gases Guideline
5. Guideline on General Principles of Process Validation
6. Guideline on Sterile Drug Products Produced by Aseptic Processing
7. Guide to Inspection of Computerized System in Drug Process
8. Regulatory Procedures Manual, Part 8
9. Inspection Operations Manual (IOM)
10. Guide to Inspections of Dosage Form Drug Manufacturers-CGMPs
11. Guide to Inspections of Lyophilization of Parenterals
12. Guide to Inspections of Pharmaceutical Quality Control Laboratories
13. Guide to Inspections of High Purity Water Systems
14. Guide to Inspections of Validation of Cleaning Processes
15. Guidance to Industry (Draft) Investigation of OOS Test Results
16. 21 CFR Part 11 Electronic Records: Electronic Signatures
17. Compliance Policy Guide7153.17 Enforcement Policy, Part 11 Electronic Records; Electronic Signatures
18. Electronic Records Guide (4/21/98) (Electronic Signatures, 21 CFR Part 11 Answers to Frequently Asked Questions)

Attachments

Attachments to the Drug Process Inspection program may be issued for certain industries, dosage forms, and processes with known problems or unique drug processes. These attachments will contain the guidance needed to perform these specialized inspections.

Some of the attachments to be issued with this program may include reporting requirements specifically designed to obtain industry-wide information on certain practices to permit evaluation of the adequacy of FDA's regulatory efforts.

Attachments and/or reporting requirements will be periodically reviewed and evaluated and deleted from the program when they are no longer needed.

CONTACTS

ORA

ORO/Division of Emergency and Investigational Operations/Drug Group (HFC-132)
Telephone: (301) 827-5653

ORO/Division of Field Science (HFC-140)
Telephone: (301) 827-7605

Center for Drug Evaluation and Research

CGMP Questions

Division of Manufacturing & Product Quality
Chief, Case Management and Guidance Branch (HFD-325)
Telephone: (301) 594-0098
Fax: (301) 594-2202

Product Quality Problems (NDA/ANDA/Compendial Specifications deviations)

Division of Manufacturing & Product Quality
Chief, Case Management & Guidance Branch (HFD-325)
Telephone: (301) 594-0098
Fax: (301) 594-2202

Product Quality Problems (Drug Product Sampling, FAR reporting, ADE reporting)

Team Leader, Post Market Surveillance Team (HFD-336)
Division of Rx Drug Compliance and Surveillance
Telephone: (301) 594-0101
Fax: (301) 594-1146

PART VII — CENTER RESPONSIBILITIES

CENTER FOR DRUG EVALUATION AND RESEARCH

The Division of Manufacturing and Product Quality (DMPQ) HFD-320 will conduct an annual evaluation in order to assess and report on the effectiveness of this program.

Please send any comments on the operation and efficiency and direct questions regarding application of the program to the Chief, Case Management and Guidance Branch, DMPQ, HFD-325, 301-594-0098, fax 301-594-2202.

APPENDIX V

21 CODE OF U.S. FEDERAL REGULATIONS PARTS 210 AND 211 CURRENT GOOD MANUFACTURING PRACTICE REGULATIONS

The Current Good Manufacturing Practice (CGMP) Regulations, 21 Code of Federal Regulations Part 210 and 211, are U.S. government regulations that describe the minimum current good manufacturing practices for the preparation of drugs and drug products for administration to humans and animals. They include methods to be used in and the facilities or controls to be used for the manufacturing, processing, packing, or holding of a drug to assure that such drug meets the requirements of the U.S. Food Drug and Cosmetic act as to safety, and has the identity and strength and meets the quality and purity characteristics that it purports or is represented to possess.

For many end users, the CGMPs may seem vague and do not provide details and direction on how one is supposed to implement the law in their own working environment. This requires drug manufactures to interact with FDA, interact with others within the industry, and continuously evaluate their level of compliance with the current regulations and current industry practices. This was done purposely by FDA as to not set a minimum standard, thus requiring interpretation and innovation, on the part of drug manufacturer. Therefore, in practice, the CGMPs are not only composed of the written law that is codified in 21 CFR sections 210 and 211, but also compose the entire body of industry knowledge, experience, as well as FDA guidance documents and findings related to FDA inspections of facilities.

Establishing a CGMP Laboratory Audit System. By David M. Bliesner
Copyright © 2006 John Wiley & Sons, Inc.

Regardless, 21 CFR sections 210 and 211 still remain the foundation of current industry practice with respect to manufacture of drug and drug product. Moreover, within these regulations are sections specifically relating to laboratory controls and remain a document with which laboratory personnel need to maintain some familiarity.

Because of this, the CGMPs are reproduced in the following text as a reference for the reader. It should be noted, however, that the CGMPs are laws subject to change and one should therefore refer to the actual regulations from time to time in order to stay current. The current version of CGMPs is accessible through the FDA website, www.fda.gov.

[Code of Federal Regulations]
[Title 21, Volume 4]
[Revised as of April 1, 2005]
From the U.S. Government Printing Office via GPO Access
[CITE: 21CFR210.1]

PART 210-CURRENT GOOD MANUFACTURING PRACTICE IN MANUFACTURING, PROCESSING, PACKING, OR HOLDING OF DRUGS; GENERAL

Sec. 210.1   Status of current good manufacturing practice regulations.
Sec. 210.2   Applicability of current good manufacturing practice regulations.
Sec. 210.3   Definitions.

Authority: 21 U.S.C. 321, 351, 352, 355, 360b, 371, 374; 42 U.S.C. 216, 262, 263a, 264.

Source: 43 FR 45076, Sept, 29, 1978, unless otherwise noted.

Sec. 210.1 Status of current good manufacturing practice regulations.

(a) The regulations set forth in this part and in parts 211 through 226 of this chapter contain the minimum current good manufacturing practice for methods to be used in, and the facilities or controls to be used for, the manufacture, processing, packing, or holding of a drug to assure that such drug meets the requirements of the act as to safety, and has the identity and strength and meets the quality and purity characteristics that it purports or is represented to possess.

(b) The failure to comply with any regulation set forth in this part and in parts 211 through 226 of this chapter in the manufacture, processing, packing, or holding of a drug shall render such drug to be adulterated under section 501(a)(2)(B) of the act and such drug, as well as the person who is responsible for the failure to comply, shall be subject to regulatory action.

(c) Owners and operators of establishments engaged in the recovery, donor screening, testing (including donor testing), processing, storage, labeling, packaging, or distribution of human cells, tissues, and cellular and tissue-based products (HCT/Ps), as defined in Sec. 1271.3(d) of this chapter, that are drugs (subject to review under an application submitted under section 505 of the act or under a biological product license application under section 351 of the Public Health Service Act), are subject to the donor-eligibility and applicable current good tissue practice procedures set forth in part 1271 subparts C and D of this chapter, in addition to the regulations in this part and in parts 211 through 226 of this chapter. Failure to comply with any applicable regulation set forth in this part, in parts 211 through 226 of this chapter, in part 1271 subpart C of this chapter, or in part 1271 subpart D of this chapter with respect to the manufacture, processing, packing or holding of a drug, renders an HCT/P adulterated under section 501(a)(2)(B) of the act. Such HCT/P, as well as the person who is responsible for the failure to comply, is subject to regulatory action.

Sec. 210.2 Applicability of current good manufacturing practice regulations.

(a) The regulations in this part and in parts 211 through 226 of this chapter as they may pertain to a drug; in parts 600 through 680 of this chapter as they may pertain to a biological product for human use; and in part 1271 of this chapter as they are applicable to a human cell, tissue, or cellular or tissue-based product (HCT/P) that is a drug (subject to review under an application submitted under section 505 of the act or under a biological product license application under section 351 of the Public Health Service Act); shall be considered to supplement, not supersede, each other, unless the regulations explicitly provide otherwise. In the event of a conflict between applicable regulations in this part and in other parts of this chapter, the regulation specifically applicable to the drug product in question shall supersede the more general.

(b) If a person engages in only some operations subject to the regulations in this part, in parts 211 through 226 of this chapter, in parts 600 through 680 of this chapter, and in part 1271 of this chapter, and not in others, that person need only comply with those regulations applicable to the operations in which he or she is engaged.

Sec. 210.3 Definitions.

(a) The definitions and interpretations contained in section 201 of the act shall be applicable to such terms when used in this part and in parts 211 through 226 of this chapter.

(b) The following definitions of terms apply to this part and to parts 211 through 226 of this chapter.

(1) *Act* means the Federal Food, Drug, and Cosmetic Act, as amended (21 U.S.C. 301 *et seq.*).

(2) *Batch* means a specific quantity of a drug or other material that is intended to have uniform character and quality, within specified limits, and is produced according to a single manufacturing order during the same cycle of manufacture.

(3) *Component* means any ingredient intended for use in the manufacture of a drug product, including those that may not appear in such drug product.

(4) *Drug product* means a finished dosage form, for example, tablet, capsule, solution, etc., that contains an active drug ingredient generally, but not necessarily, in association with inactive ingredients. The term also includes a finished dosage form that does not contain an active ingredient but is intended to be used as a placebo.

(5) *Fiber* means any particulate contaminant with a length at least three times greater than its width.

(6) *Non-fiber-releasing filter* means any filter, which after any appropriate pretreatment such as washing or flushing, will not release fibers into the component or drug product that is being filtered. All filters composed of asbestos are deemed to be fiber-releasing filters.

(7) *Active ingredient* means any component that is intended to furnish pharmacological activity or other direct effect in the diagnosis, cure, mitigation, treatment, or prevention of disease, or to affect the structure or any function of the body of man or other animals. The term includes those components that may undergo chemical change in the manufacture of the drug product and be present in the drug product in a modified form intended to furnish the specified activity or effect.

(8) *Inactive ingredient* means any component other than an *active ingredient.*

(9) *In-process material* means any material fabricated, compounded, blended, or derived by chemical reaction that is produced for, and used in, the preparation of the drug product.

(10) *Lot* means a batch, or a specific identified portion of a batch, having uniform character and quality within specified limits; or, in the case of a drug product produced by continuous process, it is a specific identified amount produced in a unit of time or quantity in a manner that assures its having uniform character and quality within specified limits.

(11) *Lot number, control number, or batch number* means any distinctive combination of letters, numbers, or symbols, or any combination of them, from which the complete history of the manufacture, processing, packing, holding, and distribution of a batch or lot of drug product or other material can be determined.

(12) *Manufacture, processing, packing, or holding of a drug product* includes packaging and labeling operations, testing, and quality control of drug products.

(13) The term *medicated feed* means any Type B or Type C medicated feed as defined in § 558.3 of this chapter. The feed contains one or more drugs as defined in section 201(g) of the act. The manufacture of medicated feeds is subject to the requirements of part 225 of this chapter.

(14) The term *medicated premix* means a Type A medicated article as in § 558.3 of this chapter. The article contains one or more drugs as defined in section 201(g) of the act. The manufacture of medicated premixes is subject to the requirements of part 226 of this chapter.

(15) *Quality control unit* means any person or organizational element designated by the firm to be responsible for the duties relating to quality control.

(16) *Strength* means:

(i) The concentration of the drug substance (for example, weight/weight, weight/volume, or unit dose/volume basis) and/or

(ii) The potency, that is, the therapeutic activity of the drug product as indicated by appropriate laboratory tests or by adequately developed and controlled clinical data (expresses, for example, in terms of units by reference to a standard).

(17) *Theoretical yield* means the quantity that would be produced at any appropriate phase of manufacture, processing, or packing of a particular drug product, based upon the quantity of components to be used, in the absence of any loss or error in actual production.

(18) *Actual yield* means the quantity that is actually produced at any appropriate phase of manufacture, processing, or packing of a particular drug product.

(19) *Percentage of theoretical yield* means the ratio of the actual yield (at any appropriate phase of manufacture, processing, or packing of a particular drug product) to the theoretical yield (at the same phase), stated as a percentage.

(20) *Acceptance criteria* means the product specifications and acceptance/rejection criteria, such as acceptable quality level and unacceptable quality level, with an associated sampling plan, that are necessary for making a decision to accept or reject a lot of batch (or any other convenient sub-groups of manufactured units).

(21) *Representative sample* means a sample that consists of a number of units that are drawn based on rational criteria such as random sampling and intended to assure that the sample accurately portrays the material being samples.

(22) *Gang-printed labeling* means labeling derived from a sheet of material on which more than one item of label is printed.

[FR 45076, Sept. 29, 1978, as amended at 51 FR 7389, Mar.3, 1986; 58 FR 41353, Aug. 3, 1993]

PART 211-CURRENT GOOD MANUFACTURING PRACTICE FOR FINISHED PHARMACEUTICALS

Subpart A-General Provisions

Subpart B-Organization and Personnel

Subpart C-Buildings and Facilities

Subpart D-Equipment

Subpart E-Control of Components and Drug Product Containers and Closures

Subpart F-Production and Process Controls

Authority: 21 U.S.C. 321, 351, 352, 355, 360b, 371, 374; 42 U.S.C. 216, 262, 263a, 264.
Source: 43 FR 45077, Sept. 29, 1978, unless otherwise noted.

Subpart A-General Provisions

Sec. 211.1 Scope.

(a) The regulations in this part contain the minimum current good manufacturing practice for preparation of drug products for administration to humans or animals.

(b) The current good manufacturing practice regulations in this chapter as they pertain to drug products; in parts 600 through 680 of this chapter, as they pertain to drugs that are also biological products for human use; and in part 1271 of this chapter, as they are applicable to drugs that are also human cells, tissues, and cellular and tissue-based products (HCT/Ps) and that are drugs (subject to review under an application submitted under section 505 of the act or under a biological product license application under section 351 of the Public Health Service Act); supplement and do not supersede the regulations in this part unless the regulations explicitly provide otherwise. In the event of a conflict between applicable regulations in this part and in other parts of this chapter, or in parts 600 through 680 of this chapter, or in part 1271 of this chapter, the regulation specifically applicable to the drug product in question shall supersede the more general.

(c) Pending consideration of a proposed exemption, published in the Federal Register of September 29, 1978, the requirements in this part shall not be enforced for OTC drug products if the products and all their ingredients are ordinarily marketed and consumed as human foods, and which products may also fall within the legal definition of drugs by virtue of their intended use. Therefore, until further notice, regulations under part 110 of this chapter, and where applicable, parts 113 to 129 of this chapter, shall be applied in determining whether these OTC drug products that are also foods are manufactured, processed, packed, or held under current good manufacturing practice.

[43 FR 45077, Sept. 29, 1978, as amended at 62 FR 66522, Dec. 19, 1997]

Sec. 211.3 Definitions.

The definitions set forth in Sec. 210.3 of this chapter apply in this part.

Subpart B-Organization and Personnel

Sec. 211.22 Responsibilities of quality control unit.

(a) There shall be a quality control unit that shall have the responsibility and authority to approve or reject all components, drug product containers, closures, in-process materials, packaging material, labeling, and drug products, and the authority to review production records to assure that no errors have occurred or, if errors have occurred, that they have been fully investigated. The quality control unit shall be responsible for approving or rejecting drug products manufactured, processed, packed, or held under contract by another company.

(b) Adequate laboratory facilities for the testing and approval (or rejection) of components, drug product containers, closures, packaging materials, in-process materials, and drug products shall be available to the quality control unit.

(c) The quality control unit shall have the responsibility for approving or rejecting all procedures or specifications impacting on the identity, strength, quality, and purity of the drug product.

(d) The responsibilities and procedures applicable to the quality control unit shall be in writing; such written procedures shall be followed.

Sec. 211.25 Personnel qualifications.

(a) Each person engaged in the manufacture, processing, packing, or holding of a drug product shall have education, training, and experience, or any combination thereof, to enable that person to perform the assigned functions. Training shall be in the particular operations that the employee performs and in current good manufacturing practice (including the current good manufacturing practice regulations in this chapter and written procedures required by these regulations) as they relate to the employee's functions. Training in current good manufacturing practice shall be conducted by qualified individuals on a continuing basis and with sufficient frequency to assure that employees remain familiar with CGMP requirements applicable to them.

(b) Each person responsible for supervising the manufacture, processing, packing, or holding of a drug product shall have the education, training, and experience, or any combination thereof, to perform assigned functions in such a manner as to provide assurance that the drug product has the safety, identity, strength, quality, and purity that it purports or is represented to possess.

(c) There shall be an adequate number of qualified personnel to perform and supervise the manufacture, processing, packing, or holding of each drug product.

Sec. 211.28 Personnel responsibilities.

(a) Personnel engaged in the manufacture, processing, packing, or holding of a drug product shall wear clean clothing appropriate for the duties they perform. Protective apparel, such as head, face, hand, and arm coverings, shall be worn as necessary to protect drug products from contamination.

(b) Personnel shall practice good sanitation and health habits.

(c) Only personnel authorized by supervisory personnel shall enter those areas of the buildings and facilities designated as limited-access areas.

(d) Any person shown at any time (either by medical examination or supervisory observation) to have an apparent illness or open lesions that may adversely affect the safety or quality of drug products shall be excluded from direct contact with components, drug product containers, closures, in-process materials, and drug products until the condition is corrected or determined by competent medical personnel not to jeopardize the safety or quality of drug products. All personnel shall be instructed to report to supervisory personnel any health conditions that may have an adverse effect on drug products.

Sec. 211.34 Consultants.

Consultants advising on the manufacture, processing, packing, or holding of drug products shall have sufficient education, training, and experience, or any combination thereof, to advise on the subject for which they are retained. Records shall be maintained stating the name, address, and qualifications of any consultants and the type of service they provide.

Subpart C-Buildings and Facilities

Sec. 211.42 Design and construction features.

(a) Any building or buildings used in the manufacture, processing, packing, or holding of a drug product shall be of suitable size, construction and location to facilitate cleaning, maintenance, and proper operations.

(b) Any such building shall have adequate space for the orderly placement of equipment and materials to prevent mix-ups between different components, drug product containers, closures, labeling, in-process materials, or drug products, and to prevent contamination. The flow of components, drug product containers, closures, labeling, in-process materials, and drug products through the building or buildings shall be designed to prevent contamination.

(c) Operations shall be performed within specifically defined areas of adequate size. There shall be separate or defined areas or such other control systems for the firm's operations as are necessary to prevent contamination or mix-ups during the course of the following procedures:

(1) Receipt, identification, storage, and withholding from use of components, drug product containers, closures, and labeling, pending the appropriate sampling, testing, or examination by the quality control unit before release for manufacturing or packaging;

(2) Holding rejected components, drug product containers, closures, and labeling before disposition;

(3) Storage of released components, drug product containers, closures, and labeling;

(4) Storage of in-process materials;

(5) Manufacturing and processing operations;

(6) Packaging and labeling operations;

(7) Quarantine storage before release of drug products;

(8) Storage of drug products after release;

(9) Control and laboratory operations;

(10) Aseptic processing, which includes as appropriate:

(i) Floors, walls, and ceilings of smooth, hard surfaces that are easily cleanable;

(ii) Temperature and humidity controls;

(iii) An air supply filtered through high-efficiency particulate air filters under positive pressure, regardless of whether flow is laminar or nonlaminar;

(iv) A system for monitoring environmental conditions;

(v) A system for cleaning and disinfecting the room and equipment to produce aseptic conditions;

(vi) A system for maintaining any equipment used to control the aseptic conditions.

(d) Operations relating to the manufacture, processing, and packing of penicillin shall be performed in facilities separate from those used for other drug products for human use.

[43 FR 45077, Sept. 29, 1978, as amended at 60 FR 4091, Jan. 20, 1995]

Sec. 211.44 Lighting.

Adequate lighting shall be provided in all areas.

Sec. 211.46 Ventilation, air filtration, air heating and cooling.

(a) Adequate ventilation shall be provided.

(b) Equipment for adequate control over air pressure, micro-organisms, dust, humidity, and temperature shall be provided when appropriate for the manufacture, processing, packing, or holding of a drug product.

(c) Air filtration systems, including prefilters and particulate matter air filters, shall be used when appropriate on air supplies to production areas. If air is recirculated to production areas, measures shall be taken to control recirculation of dust from production. In areas where air contamination occurs during production, there shall be adequate exhaust systems or other systems adequate to control contaminants.

(d) Air-handling systems for the manufacture, processing, and packing of penicillin shall be completely separate from those for other drug products for human use.

Sec. 211.48 Plumbing.

(a) Potable water shall be supplied under continuous positive pressure in a plumbing system free of defects that could contribute contamination to any drug product. Potable water shall meet the standards prescribed in the Environmental Protection Agency's Primary Drinking Water Regulations set forth in 40 CFR part 141. Water not meeting such standards shall not be permitted in the potable water system.

(b) Drains shall be of adequate size and, where connected directly to a sewer, shall be provided with an air break or other mechanical device to prevent back-siphonage.

[43 FR 45077, Sept. 29, 1978, as amended at 48 FR 11426, Mar. 18, 1983]

Sec. 211.50 Sewage and refuse.

Sewage, trash, and other refuse in and from the building and immediate premises shall be disposed of in a safe and sanitary manner.

Sec. 211.52 Washing and toilet facilities.

Adequate washing facilities shall be provided, including hot and cold water, soap or detergent, air driers or single-service towels, and clean toilet facilities easily accessible to working areas.

Sec. 211.56 Sanitation.

(a) Any building used in the manufacture, processing, packing, or holding of a drug product shall be maintained in a clean and sanitary condition, Any such building shall be free of infestation by rodents, birds, insects, and other vermin (other than laboratory animals). Trash and organic waste matter shall be held and disposed of in a timely and sanitary manner.

(b) There shall be written procedures assigning responsibility for sanitation and describing in sufficient detail the cleaning schedules, methods, equipment, and materials to be used in cleaning the buildings and facilities; such written procedures shall be followed.

(c) There shall be written procedures for use of suitable rodenticides, insecticides, fungicides, fumigating agents, and cleaning and sanitizing agents. Such written procedures shall be designed to prevent the contamination of equipment, components, drug product containers, closures, packaging, labeling materials, or drug products and shall be followed. Rodenticides, insecticides, and fungicides shall not be used unless registered and used in accordance with the Federal Insecticide, Fungicide, and Rodenticide Act (7 U.S.C. 135).

(d) Sanitation procedures shall apply to work performed by contractors or temporary employees as well as work performed by full-time employees during the ordinary course of operations.

Sec. 211.58 Maintenance.

Any building used in the manufacture, processing, packing, or holding of a drug product shall be maintained in a good state of repair.

Subpart D-Equipment

Sec. 211.63 Equipment design, size, and location.

Equipment used in the manufacture, processing, packing, or holding of a drug product shall be of appropriate design, adequate size, and suitably located to facilitate operations for its intended use and for its cleaning and maintenance.

Sec. 211.65 Equipment construction.

(a) Equipment shall be constructed so that surfaces that contact components, in-process materials, or drug products shall not be reactive, additive, or absorptive so as to alter the safety, identity, strength, quality, or purity of the drug product beyond the official or other established requirements.

(b) Any substances required for operation, such as lubricants or coolants, shall not come into contact with components, drug product containers, closures, in-process materials, or drug products so as to alter the safety, identity, strength, quality, or purity of the drug product beyond the official or other established requirements.

Sec. 211.67 Equipment cleaning and maintenance.

(a) Equipment and utensils shall be cleaned, maintained, and sanitized at appropriate intervals to prevent malfunctions or contamination that would alter the safety, identity, strength, quality, or purity of the drug product beyond the official or other established requirements.

(b) Written procedures shall be established and followed for cleaning and maintenance of equipment, including utensils, used in the manufacture, processing, packing, or holding of a drug product. These procedures shall include, but are not necessarily limited to, the following:

(1) Assignment of responsibility for cleaning and maintaining equipment;

(2) Maintenance and cleaning schedules, including, where appropriate, sanitizing schedules;

(3) A description in sufficient detail of the methods, equipment, and materials used in cleaning and maintenance operations, and the methods of disassembling and reassembling equipment as necessary to assure proper cleaning and maintenance;

(4) Removal or obliteration of previous batch identification;

(5) Protection of clean equipment from contamination prior to use;

(6) Inspection of equipment for cleanliness immediately before use.

(c) Records shall be kept of maintenance, cleaning, sanitizing, and inspection as specified in Sec. 211.180 and 211.182.

Sec. 211.68 Automatic, mechanical, and electronic equipment.

(a) Automatic, mechanical, or electronic equipment or other types of equipment, including computers, or related systems that will perform a function satisfactorily, may be used in the manufacture, processing, packing, and holding of a drug product. If such equipment is so used, it shall be routinely calibrated, inspected, or checked according to a written program designed to assure proper performance. Written records of those calibration checks and inspections shall be maintained.

(b) Appropriate controls shall be exercised over computer or related systems to assure that changes in master production and control records or other records are instituted only by authorized personnel. Input to and output from the computer or related system of formulas or other records or data shall be checked for accuracy. The degree and frequency of input/output verification shall be based on the complexity and reliability of the computer or related system. A backup file of data entered into the computer or related system shall be maintained except where certain data, such as calculations performed in connection with laboratory analysis, are eliminated by computerization or other automated processes. In such instances a written record of the program shall be maintained along with appropriate validation data. Hard copy or alternative systems, such as duplicates, tapes, or microfilm, designed to assure that backup data are exact and complete and that it is secure from alteration, inadvertent erasures, or loss shall be maintained.

[43 FR 45077, Sept. 29, 1978, as amended at 60 FR 4091, Jan. 20, 1995]

Sec. 211.72 Filters.

Filters for liquid filtration used in the manufacture, processing, or packing of injectable drug products intended for human use shall not release fibers into such products. Fiber-releasing filters may not be used in the manufacture, processing, or packing of these injectable drug products unless it is not possible to manufacture such drug products without the use of such filters. If use of a fiber-releasing filter is necessary, an additional non-fiber-releasing filter of 0.22 micron maximum mean porosity (0.45 micron if the manufacturing conditions so dictate) shall subsequently be used to reduce the content of particles in the injectable drug product. Use of an asbestos-containing filter, with or without subsequent use of a specific non-fiber-releasing filter, is permissible only upon submission of proof to the appropriate bureau of the Food and Drug Administration that use of a non-fiber-releasing filter will, or is likely to, compromise the safety or effectiveness of the injectable drug product.

Subpart E-Control of Components and Drug Product Containers and Closures

Sec. 211.80 General requirements.

(a) There shall be written procedures describing in sufficient detail the receipt, identification, storage, handling, sampling, testing, and approval or rejection of components and drug product containers and closures; such written procedures shall be followed.

(b) Components and drug product containers and closures shall at all times be handled and stored in a manner to prevent contamination.

(c) Bagged or boxed components of drug product containers, or closures shall be stored off the floor and suitably spaced to permit cleaning and inspection.

(d) Each container or grouping of containers for components or drug product containers, or closures shall be identified with a distinctive code for each lot in each shipment received. This code shall be used in recording the disposition of each lot. Each lot shall be appropriately identified as to its status (i.e., quarantine, approved, or rejected).

Sec. 211.82 Receipt and storage of untested components, drug product containers, and closures.

(a) Upon receipt and before acceptance, each container or grouping of containers of components, drug product containers, and closures shall be examined visually for appropriate labeling as to contents, container damage or broken seals, and contamination.

(b) Components, drug product containers, and closures shall be stored under quarantine until they have been tested or examined, as appropriate, and released. Storage within the area shall conform to the requirements of Sec. 211.80.

Sec. 211.84 Testing and approval or rejection of components, drug product containers, and closures.

(a) Each lot of components, drug product containers, and closures shall be withheld from use until the lot has been sampled, tested, or examined, as appropriate, and released for use by the quality control unit.

(b) Representative samples of each shipment of each lot shall be collected for testing or examination. The number of containers to be sampled, and the amount of material to be taken from each container, shall be based upon appropriate criteria such as statistical criteria for component variability, confidence levels, and degree of precision desired, the past quality history of the supplier, and the quantity needed for analysis and reserve where required by Sec. 211.170.

(c) Samples shall be collected in accordance with the following procedures:

(1) The containers of components selected shall be cleaned where necessary, by appropriate means.

(2) The containers shall be opened, sampled, and resealed in a manner designed to prevent contamination of their contents and contamination of other components, drug product containers, or closures.

(3) Sterile equipment and aseptic sampling techniques shall be used when necessary.

(4) If it is necessary to sample a component from the top, middle, and bottom of its container, such sample subdivisions shall not be composited for testing.

(5) Sample containers shall be identified so that the following information can be determined: name of the material sampled, the lot number, the container from which the sample was taken, the date on which the sample was taken, and the name of the person who collected the sample.

(6) Containers from which samples have been taken shall be marked to show that samples have been removed from them.

(d) Samples shall be examined and tested as follows:

(1) At least one test shall be conducted to verify the identity of each component of a drug product. Specific identity tests, if they exist, shall be used.

(2) Each component shall be tested for conformity with all appropriate written specifications for purity, strength, and quality. In lieu of such testing by the manufacturer, a report of analysis may be accepted from the supplier of a component, provided that at least one specific identity test is conducted on such component by the manufacturer, and provided that the manufacturer establishes the reliability of the supplier's analyses through appropriate validation of the supplier's test results at appropriate intervals.

(3) Containers and closures shall be tested for conformance with all appropriate written procedures. In lieu of such testing by the manufacturer, a certificate of testing may be accepted from the supplier, provided that at least a visual identification is conducted on such containers/closures by the manufacturer and provided that the manufacturer establishes the reliability of the supplier's test results through appropriate validation of the supplier's test results at appropriate intervals.

(4) When appropriate, components shall be microscopically examined.

(5) Each lot of a component, drug product container, or closure that is liable to contamination with filth, insect infestation, or other extraneous adulterant shall be examined against established specifications for such contamination.

(6) Each lot of a component, drug product container, or closure that is liable to microbiological contamination that is objectionable in view of its intended use shall be subjected to microbiological tests before use.

(e) Any lot of components, drug product containers, or closures that meets the appropriate written specifications of identity, strength, quality, and purity and related tests under paragraph (d) of this section may be approved and released for use. Any lot of such material that does not meet such specifications shall be rejected.

[43 FR 45077, Sept. 29, 1978, as amended at 63 FR 14356, Mar. 25, 1998]

Sec. 211.86 Use of approved components, drug product containers, and closures.

Components, drug product containers, and closures approved for use shall be rotated so that the oldest approved stock is used first. Deviation from this requirement is permitted if such deviation is temporary and appropriate.

Sec. 211.87 Retesting of approved components, drug product containers, and closures.

Components, drug product containers, and closures shall be retested or reexamined, as appropriate, for identity, strength, quality, and purity and approved or rejected by the quality control unit in accordance with Sec. 211.84 as necessary, e.g., after storage for long periods or after exposure to air, heat or other conditions that might adversely affect the component, drug product container, or closure.

Sec. 211.89 Rejected components, drug product containers, and closures.

Rejected components, drug product containers, and closures shall be identified and controlled under a quarantine system designed to prevent their use in manufacturing or processing operations for which they are unsuitable.

Sec. 211.94 Drug product containers and closures.

(a) Drug product containers and closures shall not be reactive, additive, or absorptive so as to alter the safety, identity, strength, quality, or purity of the drug beyond the official or established requirements.

(b) Container closure systems shall provide adequate protection against foreseeable external factors in storage and use that can cause deterioration or contamination of the drug product.

(c) Drug product containers and closures shall be clean and, where indicated by the nature of the drug, sterilized and processed to remove pyrogenic properties to assure that they are suitable for their intended use.

(d) Standards or specifications, methods of testing, and, where indicated, methods of cleaning, sterilizing, and processing to remove pyrogenic properties shall be written and followed for drug product containers and closures.

Subpart F-Production and Process Controls

Sec. 211.100 Written procedures; deviations.

(a) There shall be written procedures for production and process control designed to assure that the drug products have the identity, strength, quality, and purity they purport or are represented to possess. Such procedures shall include all requirements in this subpart. These written procedures, including any changes, shall be drafted, reviewed, and approved by the appropriate organizational units and reviewed and approved by the quality control unit.

(b) Written production and process control procedures shall be followed in the execution of the various production and process control functions and shall be documented at the time of performance. Any deviation from the written procedures shall be recorded and justified.

Sec. 211.101 Charge-in of components.

Written production and control procedures shall include the following, which are designed to assure that the drug products produced have the identity, strength, quality, and purity they purport or are represented to possess:

(a) The batch shall be formulated with the intent to provide not less than 100 percent of the labeled or established amount of active ingredient.

(b) Components for drug product manufacturing shall be weighed, measured, or subdivided as appropriate. If a component is removed from the original container to another, the new container shall be identified with the following information:

(1) Component name or item code;

(2) Receiving or control number;

(3) Weight or measure in new container;

(4) Batch for which component was dispensed, including its product name, strength, and lot number.

(c) Weighing, measuring, or subdividing operations for components shall be adequately supervised. Each container of component dispensed to manufacturing shall be examined by a second person to assure that:

(1) The component was released by the quality control unit;

(2) The weight or measure is correct as stated in the batch production records;

(3) The containers are properly identified.

(d) Each component shall be added to the batch by one person and verified by a second person.

Sec. 211.103 Calculation of yield.

Actual yields and percentages of theoretical yield shall be determined at the conclusion of each appropriate phase of manufacturing, processing, packaging, or holding of the drug product. Such calculations shall be performed by one person and independently verified by a second person.

Sec. 211.105 Equipment identification.

(a) All compounding and storage containers, processing lines, and major equipment used during the production of a batch of a drug product shall be properly identified at all times to indicate their contents and, when necessary, the phase of processing of the batch.

(b) Major equipment shall be identified by a distinctive identification number or code that shall be recorded in the batch production record to show the specific equipment used in the manufacture of each batch of a drug product. In cases where only one of a particular type of equipment exists in a manufacturing facility, the name of the equipment may be used in lieu of a distinctive identification number or code.

Sec. 211.110 Sampling and testing of in-process materials and drug products.

(a) To assure batch uniformity and integrity of drug products, written procedures shall be established and followed that describe the in-process controls, and tests, or examinations to be conducted on appropriate samples of in-process materials of each batch. Such control procedures shall be established to monitor the output and to validate the performance of those manufacturing processes that may be responsible for causing variability in the characteristics of in-process material and the drug product. Such control procedures shall include, but are not limited to, the following, where appropriate:

(1) Tablet or capsule weight variation;

(2) Disintegration time;

(3) Adequacy of mixing to assure uniformity and homogeneity;

(4) Dissolution time and rate;

(5) Clarity, completeness, or pH of solutions.

(b) Valid in-process specifications for such characteristics shall be consistent with drug product final specifications and shall be derived from previous acceptable process average and process variability estimates where possible and determined by the application of suitable statistical procedures where appropriate. Examination and testing of samples shall assure that the drug product and in-process material conform to specifications.

(c) In-process materials shall be tested for identity, strength, quality, and purity as appropriate, and approved or rejected by the quality control unit, during the production process, e.g., at commencement or completion of significant phases or after storage for long periods.

(d) Rejected in-process materials shall be identified and controlled under a quarantine system designed to prevent their use in manufacturing or processing operations for which they are unsuitable.

Sec. 211.111 Time limitations on production.

When appropriate, time limits for the completion of each phase of production shall be established to assure the quality of the drug product. Deviation from established time limits may be acceptable if such deviation does not compromise the quality of the drug product. Such deviation shall be justified and documented.

Sec. 211.113 Control of microbiological contamination.

(a) Appropriate written procedures, designed to prevent objectionable microorganisms in drug products not required to be sterile, shall be established and followed.

(b) Appropriate written procedures, designed to prevent microbiological contamination of drug products purporting to be sterile, shall be established and followed. Such procedures shall include validation of any sterilization process.

Sec. 211.115 Reprocessing.

(a) Written procedures shall be established and followed prescribing a system for reprocessing batches that do not conform to standards or specifications and the steps to be taken to insure that the reprocessed batches will conform with all established standards, specifications, and characteristics.

(b) Reprocessing shall not be performed without the review and approval of the quality control unit.

Subpart G-Packaging and Labeling Control

Sec. 211.122 Materials examination and usage criteria.

(a) There shall be written procedures describing in sufficient detail the receipt, identification, storage, handling, sampling, examination, and/or testing of labeling and packaging materials; such written procedures shall be followed. Labeling and packaging materials shall be representatively sampled, and examined or tested upon receipt and before use in packaging or labeling of a drug product.

(b) Any labeling or packaging materials meeting appropriate written specifications may be approved and released for use. Any labeling or packaging materials that do not meet such specifications shall be rejected to prevent their use in operations for which they are unsuitable.

(c) Records shall be maintained for each shipment received of each different labeling and packaging material indicating receipt, examination or testing, and whether accepted or rejected.

(d) Labels and other labeling materials for each different drug product, strength, dosage form, or quantity of contents shall be stored separately with suitable identification. Access to the storage area shall be limited to authorized personnel.

(e) Obsolete and outdated labels, labeling, and other packaging materials shall be destroyed.

(f) Use of gang-printed labeling for different drug products, or different strengths or net contents of the same drug product, is prohibited unless the labeling from gang-printed sheets is adequately differentiated by size, shape, or color.

(g) If cut labeling is used, packaging and labeling operations shall include one of the following special control procedures:

(1) Dedication of labeling and packaging lines to each different strength of each different drug product;

(2) Use of appropriate electronic or electromechanical equipment to conduct a 100-percent examination for correct labeling during or after completion of finishing operations; or

(3) Use of visual inspection to conduct a 100-percent examination for correct labeling during or after completion of finishing operations for hand-applied labeling. Such examination shall be performed by one person and independently verified by a second person.

(h) Printing devices on, or associated with, manufacturing lines used to imprint labeling upon the drug product unit label or case shall be monitored to assure that all imprinting conforms to the print specified in the batch production record.

[43 FR 45077, Sept. 29, 1978, as amended at 58 FR 41353, Aug. 3, 1993]

Sec. 211.125 Labeling issuance.

(a) Strict control shall be exercised over labeling issued for use in drug product labeling operations.

(b) Labeling materials issued for a batch shall be carefully examined for identity and conformity to the labeling specified in the master or batch production records.

(c) Procedures shall be used to reconcile the quantities of labeling issued, used, and returned, and shall require evaluation of discrepancies found between the quantity of drug product finished and the quantity of labeling issued when such discrepancies are outside narrow preset limits based on historical operating data. Such discrepancies shall be investigated in accordance with Sec. 211.192. Labeling reconciliation is waived for cut or roll labeling if a 100-percent examination for correct labeling is performed in accordance with Sec. 211.122(g)(2).

(d) All excess labeling bearing lot or control numbers shall be destroyed.

(e) Returned labeling shall be maintained and stored in a manner to prevent mix ups and provide proper identification.

(f) Procedures shall be written describing in sufficient detail the control procedures employed for the issuance of labeling; such written procedures shall be followed.

[43 FR 45077, Sept. 29, 1978, as amended at 58 FR 41354, Aug. 3, 1993]

Sec. 211.130 Packaging and labeling operations.

There shall be written procedures designed to assure that correct labels, labeling, and packaging materials are used for drug products; such written procedures shall be followed. These procedures shall incorporate the following features:

(a) Prevention of mix ups and cross-contamination by physical or spatial separation from operations on other drug products.

(b) Identification and handling of filled drug product containers that are set aside and held in unlabeled condition for future labeling operations to preclude mislabeling of individual containers, lots, or portions of lots. Identification need not be applied to each individual container but shall be sufficient to determine name, strength, quantity of contents, and lot or control number of each container.

(c) Identification of the drug product with a lot or control number that permits determination of the history of the manufacture and control of the batch.

(d) Examination of packaging and labeling materials for suitability and correctness before packaging operations, and documentation of such examination in the batch production record.

(e) Inspection of the packaging and labeling facilities immediately before use to assure that all drug products have been removed from previous operations. Inspection shall also be made to assure that packaging and labeling materials not suitable for subsequent operations have been removed. Results of inspection shall be documented in the batch production records.

[43 FR 45077, Sept. 29, 1978, as amended at 58 FR 41354, Aug. 3, 1993]

Sec. 211.132 Tamper-evident packaging requirements for over-the-counter (OTC) human drug products.

(a) General. The Food and Drug Administration has the authority under the Federal Food, Drug, and Cosmetic Act (the act) to establish a uniform national requirement for tamper-evident packaging of OTC drug products that will improve the security of OTC drug packaging and help assure the safety and effectiveness of OTC drug products. An OTC drug product (except a dermatological, dentifrice, insulin, or lozenge product) for retail sale that is not packaged in a tamper-resistant package or that is not properly labeled under this section is adulterated under section 501 of the act or misbranded under section 502 of the act, or both.

(b) Requirements for tamper-evident package. (1) Each manufacturer and packer who packages an OTC drug product (except a dermatological, dentifrice, insulin, or lozenge product) for retail sale shall package the product in a tamper-evident package, if this product is accessible to the public while held for sale. A tamper-evident package is one having one or more indicators or barriers to entry which, if breached or missing, can reasonably be expected to provide visible evidence to consumers that tampering has occurred. To reduce the likelihood of successful tampering and to increase the likelihood that consumers will discover if a product has been tampered with, the package is required to be distinctive by design or by the use of one or more indicators or barriers to entry that employ an identifying characteristic (e.g., a pattern, name, registered trademark, logo, or picture). For purposes of this section, the term "distinctive by design" means the packaging cannot be duplicated with commonly available materials or through commonly available processes. A tamper-evident package may involve an immediate-container and closure system or secondary-container or carton system or any combination of systems intended to provide a visual indication of package integrity. The tamper-evident feature shall be designed to and shall remain intact when handled in a reasonable manner during manufacture, distribution, and retail display.

(2) In addition to the tamper-evident packaging feature described in paragraph (b)(1) of this section, any two-piece, hard gelatin capsule covered by this section must be sealed using an acceptable tamper-evident technology.

(c) Labeling. (1) In order to alert consumers to the specific tamper-evident feature(s) used, each retail package of an OTC drug product covered by this section (except ammonia inhalant in crushable glass ampules, containers of compressed medical oxygen, or aerosol products that depend upon the power of a liquefied or compressed gas to expel the contents from the container) is required to bear a statement that:

(i) Identifies all tamper-evident feature(s) and any capsule sealing technologies used to comply with paragraph (b) of this section;

(ii) Is prominently placed on the package; and

(iii) Is so placed that it will be unaffected if the tamper-evident feature of the package is breached or missing.

(2) If the tamper-evident feature chosen to meet the requirements in paragraph (b) of this section uses an identifying characteristic, that characteristic is required to be referred to in the labeling statement. For example, the labeling statement on a bottle with a

shrink band could say "For your protection, this bottle has an imprinted seal around the neck."

(d) Request for exemptions from packaging and labeling requirements. A manufacturer or packer may request an exemption from the packaging and labeling requirements of this section. A request for an exemption is required to be submitted in the form of a citizen petition under Sec. 10.30 of this chapter and should be clearly identified on the envelope as a "Request for Exemption from the Tamper-Evident Packaging Rule."

The petition is required to contain the following:

(1) The name of the drug product or, if the petition seeks an exemption for a drug class, the name of the drug class, and a list of products within that class.

(2) The reasons that the drug product's compliance with the tamper-evident packaging or labeling requirements of this section is unnecessary or cannot be achieved.

(3) A description of alternative steps that are available, or that the petitioner has already taken, to reduce the likelihood that the product or drug class will be the subject of malicious adulteration.

(4) Other information justifying an exemption.

(e) OTC drug products subject to approved new drug applications.

Holders of approved new drug applications for OTC drug products are required under Sec. 314.70 of this chapter to provide the agency with notification of changes in packaging and labeling to comply with the requirements of this section. Changes in packaging and labeling required by this regulation may be made before FDA approval, as provided under Sec. 314.70(c) of this chapter. Manufacturing changes by which capsules are to be sealed require prior FDA approval under Sec. 314.70(b) of this chapter.

(f) Poison Prevention Packaging Act of 1970. This section does not affect any requirements for "special packaging" as defined under Sec. 310.3(l) of this chapter and required under the Poison Prevention Packaging Act of 1970.

(Approved by the Office of Management and Budget under OMB control number 0910-0149)
[54 FR 5228, Feb. 2, 1989, as amended at 63 FR 59470, Nov. 4, 1998]

Sec. 211.134 Drug product inspection.

(a) Packaged and labeled products shall be examined during finishing operations to provide assurance that containers and packages in the lot have the correct label.

(b) A representative sample of units shall be collected at the completion of finishing operations and shall be visually examined for correct labeling.

(c) Results of these examinations shall be recorded in the batch production or control records.

Sec. 211.137 Expiration dating.

(a) To assure that a drug product meets applicable standards of identity, strength, quality, and purity at the time of use, it shall bear an expiration date determined by appropriate stability testing described in Sec. 211.166.

(b) Expiration dates shall be related to any storage conditions stated on the labeling, as determined by stability studies described in Sec. 211.166.

(c) If the drug product is to be reconstituted at the time of dispensing, its labeling shall bear expiration information for both the reconstituted and unreconstituted drug products.

(d) Expiration dates shall appear on labeling in accordance with the requirements of Sec. 201.17 of this chapter.

(e) Homeopathic drug products shall be exempt from the requirements of this section.

(f) Allergenic extracts that are labeled "No U.S. Standard of Potency" are exempt from the requirements of this section.

(g) New drug products for investigational use are exempt from the requirements of this section, provided that they meet appropriate standards or specifications as demonstrated by stability studies during their use in clinical investigations. Where new drug products for investigational use are to be reconstituted at the time of dispensing, their labeling shall bear expiration information for the reconstituted drug product.

(h) Pending consideration of a proposed exemption, published in the Federal Register of September 29, 1978, the requirements in this section shall not be enforced for human OTC drug products if their labeling does not bear dosage limitations and they are stable for at least 3 years as supported by appropriate stability data.

[43 FR 45077, Sept. 29, 1978, as amended at 46 FR 56412, Nov. 17, 1981; 60 FR 4091, Jan. 20, 1995]

Subpart H-Holding and Distribution

Sec. 211.142 Warehousing procedures.

Written procedures describing the warehousing of drug products shall be established and followed. They shall include:

(a) Quarantine of drug products before release by the quality control unit.

(b) Storage of drug products under appropriate conditions of temperature, humidity, and light so that the identity, strength, quality, and purity of the drug products are not affected.

Sec. 211.150 Distribution procedures.

Written procedures shall be established, and followed, describing the distribution of drug products. They shall include:

(a) A procedure whereby the oldest approved stock of a drug product is distributed first. Deviation from this requirement is permitted if such deviation is temporary and appropriate.

(b) A system by which the distribution of each lot of drug product can be readily determined to facilitate its recall if necessary.

Subpart I-Laboratory Controls

Sec. 211.160 General requirements.

(a) The establishment of any specifications, standards, sampling plans, test procedures, or other laboratory control mechanisms required by this subpart, including any change in such specifications, standards, sampling plans, test procedures, or other laboratory control mechanisms, shall be drafted by the appropriate organizational unit and reviewed and approved by the quality control unit. The requirements in this subpart shall be followed and shall be documented at the time of performance. Any deviation from the written specifications, standards, sampling plans, test procedures, or other laboratory control mechanisms shall be recorded and justified.

(b) Laboratory controls shall include the establishment of scientifically sound and appropriate specifications, standards, sampling plans, and test procedures designed to assure that components, drug product containers, closures, in-process materials, labeling, and drug products conform to appropriate standards of identity, strength, quality, and purity. Laboratory controls shall include:

(1) Determination of conformance to appropriate written specifications for the acceptance of each lot within each shipment of components, drug product containers, closures, and labeling used in the manufacture, processing, packing, or holding of drug products. The specifications shall include a description of the sampling and testing procedures used. Samples shall be representative and adequately identified. Such procedures shall also require appropriate retesting of any component, drug product container, or closure that is subject to deterioration.

(2) Determination of conformance to written specifications and a description of sampling and testing procedures for in-process materials. Such samples shall be representative and properly identified.

(3) Determination of conformance to written descriptions of sampling procedures and appropriate specifications for drug products. Such samples shall be representative and properly identified.

(4) The calibration of instruments, apparatus, gauges, and recording devices at suitable intervals in accordance with an established written program containing specific directions, schedules, limits for accuracy and precision, and provisions for remedial action in the event accuracy and/or precision limits are not met. Instruments, apparatus, gauges, and recording devices not meeting established specifications shall not be used.

Sec. 211.165 Testing and release for distribution.

(a) For each batch of drug product, there shall be appropriate laboratory determination of satisfactory conformance to final specifications for the drug product, including the identity and strength of each active ingredient, prior to release. Where sterility and/or pyrogen testing are conducted on specific batches of short-lived radiopharmaceuticals, such batches may be released prior to completion of sterility and/or pyrogen testing, provided such testing is completed as soon as possible.

(b) There shall be appropriate laboratory testing, as necessary, of each batch of drug product required to be free of objectionable microorganisms.

(c) Any sampling and testing plans shall be described in written procedures that shall include the method of sampling and the number of units per batch to be tested; such written procedure shall be followed.

(d) Acceptance criteria for the sampling and testing conducted by the quality control unit shall be adequate to assure that batches of drug products meet each appropriate specification and appropriate statistical quality control criteria as a condition for their approval and release. The statistical quality control criteria shall include appropriate acceptance levels and/or appropriate rejection levels.

(e) The accuracy, sensitivity, specificity, and reproducibility of test methods employed by the firm shall be established and documented. Such validation and documentation may be accomplished in accordance with Sec. 211.194(a)(2).

(f) Drug products failing to meet established standards or specifications and any other relevant quality control criteria shall be rejected. Reprocessing may be performed. Prior to acceptance and use, reprocessed material must meet appropriate standards, specifications, and any other relevant criteria.

Sec. 211.166 Stability testing.

(a) There shall be a written testing program designed to assess the stability characteristics of drug products. The results of such stability testing shall be used in determining appropriate storage conditions and expiration dates. The written program shall be followed and shall include:

(1) Sample size and test intervals based on statistical criteria for each attribute examined to assure valid estimates of stability;

(2) Storage conditions for samples retained for testing;

(3) Reliable, meaningful, and specific test methods;

(4) Testing of the drug product in the same container-closure system as that in which the drug product is marketed;

(5) Testing of drug products for reconstitution at the time of dispensing (as directed in the labeling) as well as after they are reconstituted.

(b) An adequate number of batches of each drug product shall be tested to determine an appropriate expiration date and a record of such data shall be maintained. Accelerated studies, combined with basic stability information on the components, drug products, and container-closure system, may be used to support tentative expiration dates provided full shelf life studies are not available and are being conducted. Where data from accelerated studies are used to project a tentative expiration date that is beyond a date supported by actual shelf life studies, there must be stability studies conducted, including drug product testing at appropriate intervals, until the tentative expiration date is verified or the appropriate expiration date determined.

(c) For homeopathic drug products, the requirements of this section are as follows:

(1) There shall be a written assessment of stability based at least on testing or examination of the drug product for compatibility of the ingredients, and based on marketing experience with the drug product to indicate that there is no degradation of the product for the normal or expected period of use.

(2) Evaluation of stability shall be based on the same container-closure system in which the drug product is being marketed.

(d) Allergenic extracts that are labeled "No U.S. Standard of Potency" are exempt from the requirements of this section.

[43 FR 45077, Sept. 29, 1978, as amended at 46 FR 56412, Nov. 17, 1981]

Sec. 211.167 Special testing requirements.

(a) For each batch of drug product purporting to be sterile and/or pyrogen-free, there shall be appropriate laboratory testing to determine conformance to such requirements. The test procedures shall be in writing and shall be followed.

(b) For each batch of ophthalmic ointment, there shall be appropriate testing to determine conformance to specifications regarding the presence of foreign particles and harsh or abrasive substances. The test procedures shall be in writing and shall be followed.

(c) For each batch of controlled-release dosage form, there shall be appropriate laboratory testing to determine conformance to the specifications for the rate of release of each active ingredient. The test procedures shall be in writing and shall be followed.

Sec. 211.170 Reserve samples.

(a) An appropriately identified reserve sample that is representative of each lot in each shipment of each active ingredient shall be retained. The reserve sample consists of at least twice the quantity necessary for all tests required to determine whether the active ingredient meets its established specifications, except for sterility and pyrogen testing. The retention time is as follows:

(1) For an active ingredient in a drug product other than those described in paragraphs (a) (2) and (3) of this section, the reserve sample shall be retained for 1 year after the expiration date of the last lot of the drug product containing the active ingredient.

(2) For an active ingredient in a radioactive drug product, except for nonradioactive reagent kits, the reserve sample shall be retained for:

(i) Three months after the expiration date of the last lot of the drug product containing the active ingredient if the expiration dating period of the drug product is 30 days or less; or

(ii) Six months after the expiration date of the last lot of the drug product containing the active ingredient if the expiration dating period of the drug product is more than 30 days.

(3) For an active ingredient in an OTC drug product that is exempt from bearing an expiration date under Sec. 211.137, the reserve sample shall be retained for 3 years after distribution of the last lot of the drug product containing the active ingredient.

(b) An appropriately identified reserve sample that is representative of each lot or batch of drug product shall be retained and stored under conditions consistent with product labeling. The reserve sample shall be stored in the same immediate container-closure system in which the drug product is marketed or in one that has essentially the same characteristics. The reserve sample consists of at least twice the quantity necessary to perform all the required tests, except those for sterility and pyrogens. Except for those for drug products described in paragraph (b)(2) of this section, reserve samples from representative sample lots or batches selected by acceptable statistical procedures shall be examined visually at least once a year for evidence of deterioration unless visual examination would affect the integrity of the reserve sample. Any evidence of reserve sample deterioration shall be investigated in accordance with Sec. 211.192.

The results of the examination shall be recorded and maintained with other stability data on the drug product. Reserve samples of compressed medical gases need not be retained. The retention time is as follows:

(1) For a drug product other than those described in paragraphs (b) (2) and (3) of this section, the reserve sample shall be retained for 1 year after the expiration date of the drug product.

(2) For a radioactive drug product, except for nonradioactive reagent kits, the reserve sample shall be retained for:

(i) Three months after the expiration date of the drug product if the expiration dating period of the drug product is 30 days or less; or

(ii) Six months after the expiration date of the drug product if the expiration dating period of the drug product is more than 30 days.

(3) For an OTC drug product that is exempt for bearing an expiration date under Sec. 211.137, the reserve sample must be retained for 3 years after the lot or batch of drug product is distributed.

[48 FR 13025, Mar. 29, 1983, as amended at 60 FR 4091, Jan. 20, 1995]

Sec. 211.173 Laboratory animals.

Animals used in testing components, in-process materials, or drug products for compliance with established specifications shall be maintained and controlled in a manner that assures their suitability for their intended use. They shall be identified, and adequate records shall be maintained showing the history of their use.

Sec. 211.176 Penicillin contamination.

If a reasonable possibility exists that a non-penicillin drug product has been exposed to cross-contamination with penicillin, the non-penicillin drug product shall be tested for the presence of penicillin. Such drug product shall not be marketed if detectable levels are found when tested according to procedures specified in 'Procedures for Detecting and Measuring Penicillin Contamination in Drugs,' which is incorporated by reference. Copies are available from the Division of Research and Testing (HFD-470), Center for Drug Evaluation and Research, Food and Drug Administration, 5100 Paint Branch Pkwy., College Park, MD . 20740, or available for inspection at the National Archives and Records Administration (NARA). For information on the availability of this material at NARA, call 202-741-6030, or go to: http://www.archives.gov/federal—register/code—of—federal—regulations/ ibr—locations.html.

[43 FR 45077, Sept. 29, 1978, as amended at 47 FR 9396, Mar. 5, 1982; 50 FR 8996, Mar. 6, 1985; 55 FR 11577, Mar. 29, 1990; 66 FR 56035, Nov. 6, 2001; 69 FR 18803, Apr. 9, 2004]

Subpart J-Records and Reports

Sec. 211.180 General requirements.

(a) Any production, control, or distribution record that is required to be maintained in compliance with this part and is specifically associated with a batch of a drug product shall be retained for at least 1 year after the expiration date of the batch or, in the case of certain OTC drug products lacking expiration dating because they meet the criteria for exemption under Sec. 211.137, 3 years after distribution of the batch.

(b) Records shall be maintained for all components, drug product containers, closures, and labeling for at least 1 year after the expiration date or, in the case of certain OTC drug products lacking expiration dating because they meet the criteria for exemption under Sec. 211.137, 3 years after distribution of the last lot of drug product incorporating the component or using the container, closure, or labeling.

(c) All records required under this part, or copies of such records, shall be readily available for authorized inspection during the retention period at the establishment where the activities described in such records occurred. These records or copies thereof shall be subject to photocopying or other means of reproduction as part of such inspection. Records that can be immediately retrieved from another location by computer or other electronic means shall be considered as meeting the requirements of this paragraph.

(d) Records required under this part may be retained either as original records or as true copies such as photocopies, microfilm, microfiche, or other accurate reproductions of the original records. Where reduction techniques, such as microfilming, are used, suitable reader and photocopying equipment shall be readily available.

(e) Written records required by this part shall be maintained so that data therein can be used for evaluating, at least annually, the quality standards of each drug product to determine the need for changes in drug product specifications or manufacturing or control procedures. Written procedures shall be established and followed for such evaluations and shall include provisions for:

(1) A review of a representative number of batches, whether approved or rejected, and, where applicable, records associated with the batch.

(2) A review of complaints, recalls, returned or salvaged drug products, and investigations conducted under Sec. 211.192 for each drug product.

(f) Procedures shall be established to assure that the responsible officials of the firm, if they are not personally involved in or immediately aware of such actions, are notified in writing of any investigations conducted under Sec. 211.198, 211.204, or 211.208 of these regulations, any recalls, reports of inspectional observations issued by the Food and Drug Administration, or any regulatory actions relating to good manufacturing practices brought by the Food and Drug Administration.

[43 FR 45077, Sept. 29, 1978, as amended at 60 FR 4091, Jan. 20, 1995]

Sec. 211.182 Equipment cleaning and use log.

A written record of major equipment cleaning, maintenance (except routine maintenance such as lubrication and adjustments), and use shall be included in individual equipment logs that show the date, time, product, and lot number of each batch processed. If equipment is dedicated to manufacture of one product, then individual equipment logs are not required, provided that lots or batches of such product follow in numerical order and are manufactured in numerical sequence. In cases where dedicated equipment is employed, the records of cleaning, maintenance, and use shall be part of the batch record. The persons performing and double-checking the cleaning and maintenance shall date and sign or initial the log indicating that the work was performed. Entries in the log shall be in chronological order.

Sec. 211.184 Component, drug product container, closure, and labeling records.

These records shall include the following:

(a) The identity and quantity of each shipment of each lot of components, drug product containers, closures, and labeling; the name of the supplier; the supplier's lot number(s) if known; the receiving code as specified in Sec. 211.80; and the date of receipt. The name and location of the prime manufacturer, if different from the supplier, shall be listed if known.

(b) The results of any test or examination performed (including those performed as required by Sec. 211.82(a), Sec. 211.84(d), or Sec. 211.122(a) and the conclusions derived therefrom.

(c) An individual inventory record of each component, drug product container, and closure and, for each component, a reconciliation of the use of each lot of such component. The inventory record shall contain sufficient information to allow determination of any batch or lot of drug product associated with the use of each component, drug product container, and closure.

(d) Documentation of the examination and review of labels and labeling for conformity with established specifications in accord with Sec. 211.122(c) and 211.130(c).

(e) The disposition of rejected components, drug product containers, closure, and labeling.

Sec. 211.186 Master production and control records.

(a) To assure uniformity from batch to batch, master production and control records for each drug product, including each batch size thereof, shall be prepared, dated, and signed (full signature, handwritten) by one person and independently checked, dated, and signed by a second person. The preparation of master production and control records shall be described in a written procedure and such written procedure shall be followed.

(b) Master production and control records shall include:

(1) The name and strength of the product and a description of the dosage form;

(2) The name and weight or measure of each active ingredient per dosage unit or per unit of weight or measure of the drug product, and a statement of the total weight or measure of any dosage unit;

(3) A complete list of components designated by names or codes sufficiently specific to indicate any special quality characteristic;

(4) An accurate statement of the weight or measure of each component, using the same weight system (metric, avoirdupois, or apothecary) for each component. Reasonable variations may be permitted, however, in the amount of components necessary for the preparation in the dosage form, provided they are justified in the master production and control records;

(5) A statement concerning any calculated excess of component;

(6) A statement of theoretical weight or measure at appropriate phases of processing;

(7) A statement of theoretical yield, including the maximum and minimum percentages of theoretical yield beyond which investigation according to Sec. 211.192 is required;

(8) A description of the drug product containers, closures, and packaging materials, including a specimen or copy of each label and all other labeling signed and dated by the person or persons responsible for approval of such labeling;

(9) Complete manufacturing and control instructions, sampling and testing procedures, specifications, special notations, and precautions to be followed.

Sec. 211.188 Batch production and control records.

Batch production and control records shall be prepared for each batch of drug product produced and shall include complete information relating to the production and control of each batch. These records shall include:

(a) An accurate reproduction of the appropriate master production or control record, checked for accuracy, dated, and signed;

(b) Documentation that each significant step in the manufacture, processing, packing, or holding of the batch was accomplished, including:

(1) Dates;

(2) Identity of individual major equipment and lines used;

(3) Specific identification of each batch of component or in-process material used;

(4) Weights and measures of components used in the course of processing;

(5) In-process and laboratory control results;

(6) Inspection of the packaging and labeling area before and after use;

(7) A statement of the actual yield and a statement of the percentage of theoretical yield at appropriate phases of processing;

(8) Complete labeling control records, including specimens or copies of all labeling used;

(9) Description of drug product containers and closures;

(10) Any sampling performed;

(11) Identification of the persons performing and directly supervising or checking each significant step in the operation;

(12) Any investigation made according to Sec. 211.192.

(13) Results of examinations made in accordance with Sec. 211.134.

Sec. 211.192 Production record review.

All drug product production and control records, including those for packaging and labeling, shall be reviewed and approved by the quality control unit to determine compliance with all established, approved written procedures before a batch is released or distributed. Any unexplained discrepancy (including a percentage of theoretical yield exceeding the maximum or minimum percentages established in master production and control records) or the failure of a batch or any of its components to meet any of its specifications shall be thoroughly investigated, whether or not the batch has already been distributed. The investigation shall extend to other batches of the same drug product and other drug products that may have been associated with the specific failure or discrepancy. A written record of the investigation shall be made and shall include the conclusions and follow-up.

Sec. 211.194 Laboratory records.

(a) Laboratory records shall include complete data derived from all tests necessary to assure compliance with established specifications and standards, including examinations and assays, as follows:

(1) A description of the sample received for testing with identification of source (that is, location from where sample was obtained), quantity, lot number or other distinctive code, date sample was taken, and date sample was received for testing.

(2) A statement of each method used in the testing of the sample. The statement shall indicate the location of data that establish that the methods used in the testing of the sample meet proper standards of accuracy and reliability as applied to the product tested. (If the method employed is in the current revision of the United States Pharmacopoeia, National Formulary, Association of Official Analytical Chemists, Book of Methods,\1\ or in other recognized standard references, or is detailed in an approved new drug application and the referenced method is not modified, a statement indicating the method and reference will suffice). The suitability of all testing methods used shall be verified under actual conditions of use.

NOTE: Copies may be obtained from: Association of Official Analytical Chemists, 2200 Wilson Blvd., Suite 400, Arlington, VA 22201-3301.

----

(3) A statement of the weight or measure of sample used for each test, where appropriate.

(4) A complete record of all data secured in the course of each test, including all graphs, charts, and spectra from laboratory instrumentation, properly identified to show the specific component, drug product container, closure, in-process material, or drug product, and lot tested.

(5) A record of all calculations performed in connection with the test, including units of measure, conversion factors, and equivalency factors.

(6) A statement of the results of tests and how the results compare with established standards of identity, strength, quality, and purity for the component, drug product container, closure, in-process material, or drug product tested.

(7) The initials or signature of the person who performs each test and the date(s) the tests were performed.

(8) The initials or signature of a second person showing that the original records have been reviewed for accuracy, completeness, and compliance with established standards.

(b) Complete records shall be maintained of any modification of an established method employed in testing. Such records shall include the reason for the modification and data to verify that the modification produced results that are at least as accurate and reliable for the material being tested as the established method.

(c) Complete records shall be maintained of any testing and standardization of laboratory reference standards, reagents, and standard solutions.

(d) Complete records shall be maintained of the periodic calibration of laboratory instruments, apparatus, gauges, and recording devices required by Sec. 211.160(b)(4).

(e) Complete records shall be maintained of all stability testing performed in accordance with Sec. 211.166.

[43 FR 45077, Sept. 29, 1978, as amended at 55 FR 11577, Mar. 29, 1990; 65 FR 18889, Apr. 10, 2000]

Sec. 211.196 Distribution records.

Distribution records shall contain the name and strength of the product and description of the dosage form, name and address of the consignee, date and quantity shipped, and lot or control number of the drug product. For compressed medical gas products, distribution records are not required to contain lot or control numbers.
(Approved by the Office of Management and Budget under control number 0910-0139)
[49 FR 9865, Mar. 16, 1984]

Sec. 211.198 Complaint files.

(a) Written procedures describing the handling of all written and oral complaints regarding a drug product shall be established and followed. Such procedures shall include provisions for review by the quality control unit, of any complaint involving the possible failure of a drug product to meet any of its specifications and, for such drug products, a determination as to the need for an investigation in accordance with Sec. 211.192. Such procedures shall include provisions for review to determine whether the complaint represents a serious and unexpected adverse drug experience which is required to be reported to the Food and Drug Administration in accordance with Sec. 310.305 and 514.80 of this chapter.

(b) A written record of each complaint shall be maintained in a file designated for drug product complaints. The file regarding such drug product complaints shall be maintained at the establishment where the drug product involved was manufactured, processed, or packed, or such file may be maintained at another facility if the written records in such files are readily available for inspection at that other facility. Written records involving a drug product shall be maintained until at least 1 year after the expiration date of the drug product, or

1 year after the date that the complaint was received, whichever is longer. In the case of certain OTC drug products lacking expiration dating because they meet the criteria for exemption under Sec. 211.137, such written records shall be maintained for 3 years after distribution of the drug product.

(1) The written record shall include the following information, where known: the name and strength of the drug product, lot number, name of complainant, nature of complaint, and reply to complainant.

(2) Where an investigation under Sec. 211.192 is conducted, the written record shall include the findings of the investigation and follow-up. The record or copy of the record of the investigation shall be maintained at the establishment where the investigation occurred in accordance with Sec. 211.180(c).

(3) Where an investigation under Sec. 211.192 is not conducted, the written record shall include the reason that an investigation was found not to be necessary and the name of the responsible person making such a determination.

[43 FR 45077, Sept. 29, 1978, as amended at 51 FR 24479, July 3, 1986; 68 FR 15364, Mar. 31, 2003]

Subpart K-Returned and Salvaged Drug Products

Sec. 211.204 Returned drug products.

Returned drug products shall be identified as such and held. If the conditions under which returned drug products have been held, stored, or shipped before or during their return, or if the condition of the drug product, its container, carton, or labeling, as a result of storage or shipping, casts doubt on the safety, identity, strength, quality or purity of the drug product, the returned drug product shall be destroyed unless examination, testing, or other investigations prove the drug product meets appropriate standards of safety, identity, strength, quality, or purity. A drug product may be reprocessed provided the subsequent drug product meets appropriate standards, specifications, and characteristics. Records of returned drug products shall be maintained and shall include the name and label potency of the drug product dosage form, lot number (or control number or batch number), reason for the return, quantity returned, date of disposition, and ultimate disposition of the returned drug product. If the reason for a drug product being returned implicates associated batches, an appropriate investigation shall be conducted in accordance with the requirements of Sec. 211.192. Procedures for the holding, testing, and reprocessing of returned drug products shall be in writing and shall be followed.

Sec. 211.208 Drug product salvaging.

Drug products that have been subjected to improper storage conditions including extremes in temperature, humidity, smoke, fumes, pressure, age, or radiation due to natural disasters, fires, accidents, or equipment failures shall not be salvaged and returned to the marketplace. Whenever there is a question whether drug products have been subjected to such conditions, salvaging operations may be conducted only if there is (a) evidence from laboratory tests and assays (including animal feeding studies where applicable) that the drug products meet all applicable standards of identity, strength, quality, and purity and (b) evidence from inspection of the premises that the drug products and their associated packaging were not subjected to improper storage conditions as a result of the disaster or accident. Organoleptic examinations shall be acceptable only as supplemental evidence that the drug products meet appropriate standards of identity, strength, quality, and purity. Records including name, lot number, and disposition shall be maintained for drug products subject to this section.

INDEX

Establishing a CGMB Laboratory Audit System, By David M. Bliesner
Copyright © 2006 John Wiley & Sons, Inc.